PRAISE FOR
SCRATCH PROGRAMMING
PLAYGROUND

"An impressive manual for achieving Scratch progra~~~ creating genuinely entertaining games."

—Kirkus Reviews

"My son was able to successfully complete each game he attempted. And along the way, he gained some excellent programming skills. I also really appreciated the character lessons he learned (patience and perseverance, not to mention goal-setting). These were a great by-product of Scratch programming!"

—The Old Schoolhouse Magazine

"The book is well written, full of humor and puns, and the explanations of how things work are good."

—*I Programmer*

"I'm quite impressed with exactly how much of the Scratch programming tool the reader will have used and learned by the end of the book, and I think teachers and parents will find this a 5-star reference."

—*Jim Kelly, GeekDad*

"If you are looking for the next thing for your Minecraft-loving kids and haven't tried Scratch, the *Scratch Programming Playground* book is a great guide to teach kids how to program by making cool games."

—*Tech Savvy Mama*

SCRATCH 3 PROGRAMMING PLAYGROUND

LEARN TO PROGRAM BY MAKING COOL GAMES

BY AL SWEIGART

no starch press

San Francisco

Printed in the United States of America

First printing

24 23 22 21   1 2 3 4 5 6 7 8 9

ISBN-13: 978-1-7185-0021-1 (print)
ISBN-13: 978-1-7185-0022-8 (ebook)

Publisher: William Pollock
Executive Editor: Barbara Yien
Production Editor: Rachel Monaghan
Developmental Editor: Frances Saux
Cover Illustration: Josh Ellingson
Illustrator: Miran Lipovača
Interior Design: Beth Middleworth
Technical Reviewer: Olivia Rodrigues
Copyeditor: Kim Wimpsett
Compositor: Happenstance Type-O-Rama
Proofreader: Paula L. Fleming

For information on book distributors or translations, please contact No Starch Press, Inc. directly:
No Starch Press, Inc.
245 8th Street, San Francisco, CA 94103
phone: 1-415-863-9900; info@nostarch.com
www.nostarch.com

The Library of Congress has cataloged the first edition as follows:

```
Names: Sweigart, Al, author.
Title: Scratch programming playground : learn to program by making cool games
   / Al Sweigart.
Description: San Francisco : No Starch Press, [2016] | Audience: Ages 8+. |
   Includes index.
Identifiers: LCCN 2016022304 (print) | LCCN 2016024406 (ebook) | ISBN
   9781593277628 | ISBN 1593277628 | ISBN 9781593277963 (epub) | ISBN
   1593277962 (epub) | ISBN 9781593277970 ( mobi) | ISBN 1593277970 (mobi)
Subjects: LCSH: Scratch (Computer program language)--Juvenile literature. |
   Computer games--Programming--Juvenile literature. | Computer
   programming--Juvenile literature. | Microcomputers--Programming--Juvenile
   literature.
Classification: LCC QA76.73.S345 S94 2016 (print) | LCC QA76.73.S345 (ebook)
   | DDC 005.13/3--dc23
LC record available at https://lccn.loc.gov/2016022304
```

For Seymour Papert,
So long and thanks for all the turtles
(February 29, 1928 – July 31, 2016)

ABOUT THE AUTHOR

Al Sweigart is a software developer, tech book author, and hoopy frood who really knows where his towel is. He has written several programming books for beginners, including *Automate the Boring Stuff with Python*, also from No Starch Press. His books are freely available under a Creative Commons license at his website *http://www.inventwithpython.com/*.

ABOUT THE TECHNICAL REVIEWER

Olivia Rodrigues is a mathematics, computing, and science teacher and the head of computing at a private elementary school in England. In addition to holding the British Computer Society certificate in Computer Science teaching, she is a Google Certified Educator, a Microsoft Innovative Educator Expert, and an IDEA Award Teacher Ambassador. She has a passion for encouraging children to learn how to code. She is the mother of a one-year-old girl called Geneviève and wife to Gary.

BRIEF CONTENTS

CONTENTS IN DETAIL

6
ASTEROID BREAKER . . . IN SPACE!   119

7
MAKING AN ADVANCED PLATFORMER   147

ACKNOWLEDGMENTS

It's misleading to have just my name on the cover. This book wouldn't exist without the efforts of many people. I'd like to thank my publisher, Bill Pollock; my editor, Frances Saux; my production editor, Rachel Monaghan; my technical reviewer, Olivia Rodrigues; and my copyeditor, Kim Wimpsett. And I'd like to thank everyone who worked on the first edition: my editors, Laurel Chun and Tyler Ortman; my technical reviewer, Martin Tan; my copyeditor, Anne Marie Walker; and all of the staff at No Starch Press.

Thanks to the MIT Media Lab's Lifelong Kindergarten group for their development of Scratch, which has a long chain of influential thinkers: Mitchel Resnick, Seymour Papert, Marvin Minsky, and Jean Piaget. While we give the younger generation a ride on our shoulders, let's never forget where we ourselves stand.

Special thanks to the Museum of Art and Digital Entertainment in Oakland, California. A video game museum is as fun to be involved with as it sounds, and volunteering with MADE's weekend Scratch class has been thoroughly rewarding. If Alex Handy, Mike Pavone, and William Morgan hadn't started the Scratch class, I never would've come up with the idea for this book.

INTRODUCTION

laying video games is fun, but programming your own video games is a creative, challenging skill that will let you make your own fun. The free Scratch programming environment gives everyone an easy way to learn programming skills. While Scratch is primarily designed for 8- to 16-year-olds, it's used by people of all ages, including younger children with their parents and college students learning their first programming language.

There's so much that you can do with Scratch, it's hard to know where to start. That's where this book comes in. This book guides you through creating several video games in Scratch. By building the projects in this book, you'll get a good idea of which blocks are commonly used to create video games in Scratch. These projects provide a solid foundation for you to build upon when creating your own original programs.

WHO THIS BOOK IS FOR

No previous programming experience is necessary to read this book. The only mathematics skills required are basic arithmetic: addition, subtraction, multiplication, and division. Don't let math phobia prevent you from learning to program. And don't forget that the computer will perform calculations for you!

Each program in the book is easy to make by following the step-by-step instructions. You'll learn about the code blocks and programming concepts as you make games that use them. No matter your skill level, there's no reason you can't start reading this book now.

Kids can follow along with the activities on their own, but this book is also for parents, teachers, and volunteers who want to introduce their children or students to the world of programming. The projects are ideal for a weekend activity or afterschool computer club. Adults don't have to be software engineers to use this book to help others learn.

If you want a thorough guide to all of Scratch's features, you can watch video tutorials online at *https://scratch.mit.edu/help/videos/* and *https://inventwithscratch.com/*. I also recommend the book *25 Scratch 3 Games* by Max Wainewright (No Starch Press, 2019) at *https://nostarch.com/25scratchgames/*.

But programming is a *hands-on* skill like karate or guitar: you can't learn it by reading alone. Make sure you're following along and creating the games—you'll learn a lot more that way.

ABOUT THIS BOOK

Each chapter walks you through programming a single game, and programming concepts are explained as they come up. You'll start by sketching out what the final game will look like and planning the main parts of the parts of the program. The sections that follow will cover how to code each of these parts step-by-step until you've built the full game. After building the main game, you'll have the option of adding special features and cheat modes. Review questions at the end of each chapter help you check whether you understand the topics covered.

Chapter 1: Getting Started with Scratch shows you how to access the Scratch website and the different parts of the Scratch editor.

In **Chapter 2: Rainbow Lines . . . in Space!**, you'll create an animated art project using basic code blocks and several sprites working together. You'll also learn about directions and degrees.

In **Chapter 3: Maze Runner**, you'll make a maze game in which the player uses the keyboard to change the cat's coordinates and guide it through eight different maze levels.

Chapter 4: Shooting Hoops with Gravity shows you how to make a basketball game that implements realistic gravity for jumping cats and falling basketballs.

Chapter 5: A Polished Brick Breaker Game covers simple techniques for taking a plain brick breaker game and turning it into a polished, exciting game with animations, sound effects, and more.

Chapter 6: Asteroid Breaker . . . in Space! features a clone of the classic space shooter *Asteroids*. You'll add mouse and keyboard controls to the spaceship.

Chapter 7: Making an Advanced Platformer pulls together many of the concepts used in previous chapters and explains how to create a platformer game with walking and jumping animations, platforms, and AI-controlled enemies.

HOW TO USE THIS BOOK

All of the book's projects start with a sketch of the game we'll make. The labels on the sketch point to features that we'll add to the game with code.

To keep things manageable, we'll tackle each game one part at a time. The blue ABC headings in the book correspond to these features in the sketch.

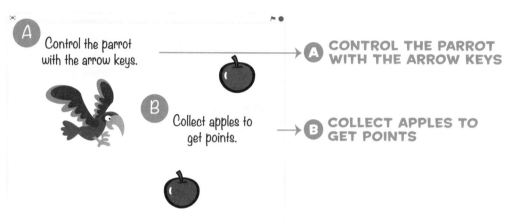

Splitting a big problem into a bunch of little problems can really help organize your thinking and make a big problem feel approachable. After we get a simple version of the game up and running, we'll add new features, cheat codes, and more. When you're ready to start creating games yourself, I recommend starting with a simple sketch.

SAVE POINT

Throughout this book, you'll see these "Save Point" boxes. Because you'll be making programs step-by-step, you'll want to pause and run the program every so often, even if it isn't finished. You'll be able to see whether the program is working correctly so far and catch any mistakes early. The Save Point boxes will also remind you to save the program by selecting **File ▶ Save to your computer** in the menu bar.

ONLINE RESOURCES

Although the Scratch environment includes
many images, you'll need some extra files
to make the projects in this book. These
files are in the resources ZIP file, which
you can download from *https://www
.nostarch.com/scratch-playground2e/*.
You'll need to unzip the files to your hard
drive to access them.

The resources ZIP file contains the
image files used in the book's projects
and the skeleton project files for each
of the programs. These skeleton project
files have all the setup steps already completed and require you only to add the
code blocks. If you are having trouble finishing your program,
you can try starting from the skeleton project file instead of a
new, blank project. Using the skeleton project files may be a
good idea if you are a teacher who is coaching several students
and time is limited, because they'll only have to add the code
blocks to complete the program.

ERRATA AND UPDATES

While we've done our best to keep this book error-free,
corrections and updates for this book will be listed at
https://nostarch.com/scratch3playground/.

1

GETTING STARTED WITH SCRATCH

cratch is the best educational programming software available today. No other tool makes programming as easy as Scratch does. Many similar products have been inspired by Scratch, but Scratch remains the most popular. With Scratch, you can create interactive games, animations, and science projects, all while having lots of fun!

Scratch is a free programming environment that runs in your web browser. It was designed by the MIT Media Lab's Lifelong Kindergarten Group. Scratch users, called *Scratchers*, can create programs by snapping together code blocks in the Scratch editor. Although Scratch was designed for 8- to 16-year-olds, Scratchers consist of people of all ages, including younger children with their parents. The software makes it easy for anyone to start developing their programming and problem-solving skills.

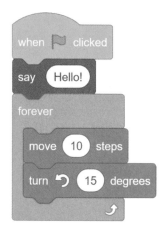

Because Scratch runs in your web browser, there's no software to install. It's impossible for a Scratch program to damage the files on your computer. Scratch is completely free—there are no ads or in-app purchases, so kids can play with everything on the Scratch site and adults don't have to worry about accidental charges.

In Scratch, you use the mouse to drag and drop code blocks, so little typing is needed. Here's an example of the snap-together code blocks:

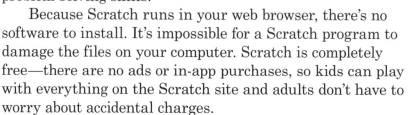

The visual Scratch editor provides you with quick feedback, so you don't have to type mysterious commands for hours before you can see your programs come to life. Scratch makes programming immediate and fun. And unlike other programming languages, Scratch doesn't have any error messages that pop up and confuse the programmer. If you want to learn the basics of programming (or help someone else learn), Scratch is second to none.

You can find answers to frequently asked questions at *https://scratch.mit.edu/info/faq/*.

RUNNING SCRATCH

To start using Scratch, open your web browser and go to *https://scratch.mit.edu/*. It doesn't matter whether you're running Windows, macOS, Linux, Android, or iOS. Scratch 3 will run on tablets, smartphones, and Raspberry Pi.

Signing up for an account is free. You can create Scratch programs without an account, but having the Scratch account lets you save your programs online. Then you can continue working on them later from any device connected to the internet.

Click the **Join Scratch** link at the top of the page to create an account. A new window opens:

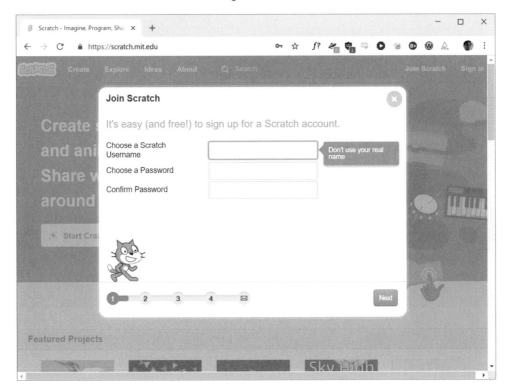

Choose a username and password and enter your account information. Scratch will never share your email address or personal information without your permission. Its full privacy policy is at *https://scratch.mit.edu/privacy_policy/*. Keep your private information safe; don't use your real name and never share your password except with parents or teachers. Don't give your password to anyone claiming to be a Scratch or MIT employee (real employees never ask for it). Don't reuse the same password you use for email or any other online account, because if one account's password gets hacked, the hacker could access your other accounts too.

After you've logged in to the Scratch website, click the **Create** link at the top of the page to start the Scratch editor.

SCRATCH DESKTOP: THE OFFLINE EDITOR

Scratch Desktop, the offline Scratch editor, lets you program without being connected to the internet. If you don't have internet access or if your Wi-Fi is unreliable, you can install the offline editor on your computer instead of using the Scratch website. The only difference is that programs will be saved on your computer instead of on the Scratch website. You can later upload your Scratch programs or copy them to a USB flash drive to move them to another computer.

The Scratch offline editor is available at *https://scratch .mit.edu/download/*.

NOTE *You may find the editor software for earlier versions, Scratch 1.4 or 2.0. Don't use these versions; they're out-of-date and don't have the new features that Scratch 3 has. If you're using Scratch in your web browser, you're using Scratch 3. If you download an offline Scratch editor, be sure to download Scratch 3.*

THE SCRATCH EDITOR AND SPRITES

The Scratch editor is where you snap code blocks together to create your game, animation, or artwork. The **Create** link at the top of the page opens the editor, as shown in the following figure, so you can start making Scratch programs:

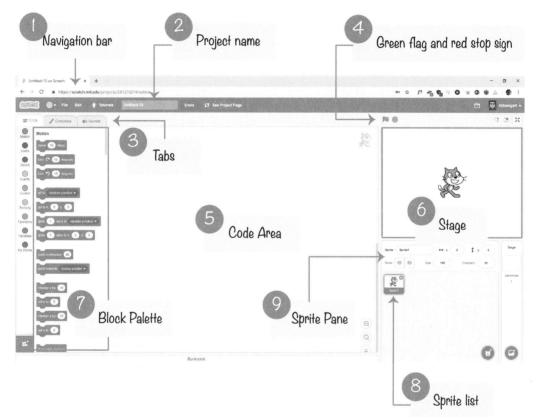

The most basic object in Scratch is the *sprite*. Sprites appear on the Stage ⑥, and their code blocks control their behavior. The editor automatically starts with a cat sprite for all new projects, but you can delete it or add more sprites. You can program a sprite by adding code blocks to the Code Area ⑤ on the right side of the screen. In Scratch, a stack of code blocks is called a *script*.

The text field at the top of the editor contains the project name ❷. After you've named your project using a descriptive name, remember to occasionally save your project by clicking **File ▶ Save Now** from the navigation bar ❶ to avoid losing your work if your browser crashes.

You access the code blocks from the Block Palette ❼ in the center. At the top of the Block Palette are nine categories of code blocks: *Motion, Looks, Sound, Events, Control, Sensing, Operators, Variables,* and *My Blocks.* Every code block belongs to one category and is the color of that category. For example, the **say** block comes from the purple *Looks* category. An infinite supply of code blocks is available; just drag them from the Block Palette to the Code Area.

Each sprite has its own scripts. When you click the sprite in the Sprite List ❽, that sprite's scripts will display in the Code Area, and its properties will display in the Sprite Pane ❾. Select the Scripts tab ❸ to display the Code Area. The Code Area is replaced by the Paint Editor and Sound Editor when the Costumes and Sounds tabs are selected, respectively.

Clicking the green flag will start your program, and clicking the red stop sign will stop it ❹.

THE PAINT EDITOR

There are a few ways to get sprites into your programs. You can use the sprites that come with Scratch, upload sprites from your computer, or draw your own. If you want to draw your own, you can use Scratch's Paint Editor.

The Paint Editor is similar to other painting programs, such as Microsoft Paint or Paintbrush. To draw a new sprite, click the **Paintbrush** button next to *New sprite.* You can change how sprites look by switching to one of many costumes. To create a new costume for a sprite, click the **Costumes** tab and then click the **Paintbrush** button next to *New costume.*

The main parts of the Paint Editor are

▶ The drawing tools, which you can select using the buttons on the left side

▶ The Canvas, where you draw images

▶ The Costume center, which indicates the center of the costume with the crosshairs symbol

▶ The Line width selector, which sets the width of the drawing tools

▶ The Color selectors, which change the color of the drawing tools

▶ The Zoom buttons for zooming into or out of the canvas

▶ The Undo and Redo buttons, which can help you correct mistakes

The Paint Editor looks like this:

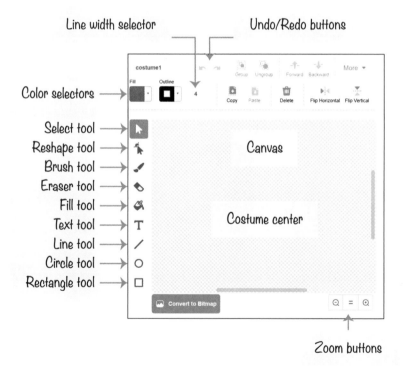

Experiment with the Paint Editor by clicking the drawing tool buttons and dragging the mouse over the Canvas to see how the tools work. Change the color and width of the drawing tools with the Color selectors and Line width selector. Then, use the Brush tool to draw some lines on the costume. If you make a mistake, click the Undo button at the top.

The list of sprite costumes is in the column to the left of the drawing tools. If you want to save a costume as an image file, right-click the costume and select **Export**.

WORKING WITH CODE BLOCKS

Before you begin programming, it's good to get an idea of how the code blocks snap together in the editor. Throughout this book, you'll learn what each code block does.

Adding Blocks

To create a new code block, drag it from the center Block Palette to the Code Area. The code blocks that have a notch on top and bump on the bottom are called *stack* blocks. To snap a stack block together with another stack block, drag the block close to the bottom of the other. When a gray background appears, drop the block to connect it to the stack.

Stack blocks can also fit in between blocks. Look carefully at where the gray background appears in the script: this is where the block will snap into place. This figure shows a **wait 1 seconds** block being moved into the middle of a script:

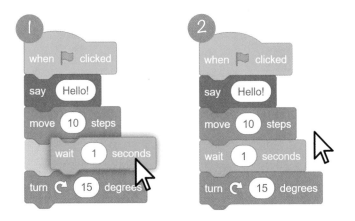

You can change a white field inside a block by clicking the white area and entering new input.

The rounded blocks are called *reporter* blocks. They fit inside the white fields. For example, in the following figure, the green **pick random 1 to 10** block fits inside the white field. When the *left edge* of the reporter block is over the white field, a white outline appears around the white field. If the left edge isn't over the white field, the white outline won't appear, and the reporter block cannot be placed inside.

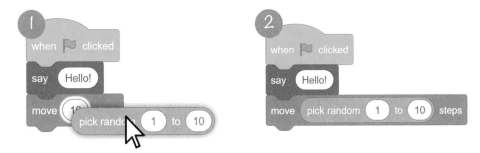

Deleting Blocks

To remove blocks, drag them out of the script. If you remove a stack block, you'll also remove the stack blocks connected beneath it, as shown in the next figure. You may need to set aside these blocks if you want to reconnect some of them to the script. Drag the blocks you want to delete over the Block Palette to remove them from the Stage. You can always add more blocks from the Block Palette when you need them.

You can also right-click a block and select **Delete Block** from the menu that appears. If you accidentally delete some blocks, you may be able to restore them by selecting **Edit ▶ Restore** from the navigation bar.

RUNNING PROGRAMS

Create the following program by dragging blocks from the Block Palette to the Code Area:

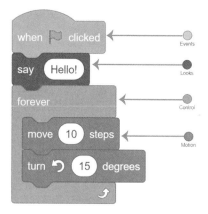

When you click the green flag at the top of the Stage, this program will start. Programs begin at the top block (**when green flag clicked**) and then run the next code block in the script. In this example, a speech bubble appears above the sprite and displays the word "Hello!" In the **forever** loop, the sprite moves forward 10 steps and then turns counterclockwise by 15 degrees. When the program gets to the last block, it *loops* back to the top. All the blocks in the **forever** block will run in a loop forever. The program stops only when you click the red stop sign.

You can also run a script or block by clicking it. But clicking the green flag is the normal way to start your program.

You can have as many sprites and code blocks in your programs as you want. As you create the programming projects in this book, you'll learn about Scratch's many different types of code blocks.

SHOWING OFF YOUR PROGRAMS

When you're logged in to your Scratch account, click the **Share** button at the top-right corner of the editor to let other Scratchers see your program. They'll be able to play your game and leave comments. If Scratchers enjoy playing the game, they can Like and Favorite your program.

Once you've finished a project, you can also add it to the Scratch Programming Playground studio. This studio features projects you and other readers have made. Once you've shared your project in Scratch, copy the URL and go to the studio's page at *https://inventwithscratch.com/studio/.* Click the **Add Projects** button, paste the URL into the text field, and click **Add by URL**. Now other readers will be able to view your game in the studio!

Don't worry if you think your game isn't good enough. Everyone begins their coding journey with simple games. Most people on the Scratch website are beginners, too. Tens of millions of people have shared their programs on the Scratch website, so don't fret if yours doesn't get many views. Games can be difficult to find with so many available on the site.

GETTING HELP

Becoming a super programmer isn't about knowing all the answers; it's knowing how to find answers. You can follow the steps for the projects in this book, but you might have questions of your own.

The Tutorials Window

At the top of the Scratch editor is the Tutorials menu button. Click this button to bring up the Tutorials window, which has links to several Scratch video tutorials.

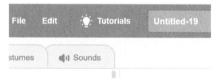

The See Inside Button

You can learn many new techniques by looking at other Scratchers' code. Find a project you like on the Scratch website and then click the **See inside** button.

Click the See inside button to view other Scratchers' code.

:tions

lgorithm for creating a candlelight flicker
iation. This is just an animation demo. but vou can

You're allowed to copy and modify, or *remix*, other Scratchers' code. All Scratch programs on the website are automatically released under a Creative Commons license, so you don't need to ask the original creator for permission, as long as you give them credit. Scratchers often remix each other's programs to create their own versions.

Still need help and want to talk to other Scratchers? Click the **Discussion Forums** link at the bottom of the home page at *https://scratch.mit.edu/* to visit the discussion forums.

SUMMARY

The Scratch editor is a creative tool with great potential. You'll see all sorts of Scratch projects on the Scratch website: games, cartoons, simulations, and informative presentation slides.

Now that you know how to access the Scratch website, create an account, use the Scratch and Paint editors, and snap together code blocks into scripts, you're ready to follow the step-by-step instructions in the rest of this book. If you want help, be sure to use the Tutorials window in the Scratch editor and the discussion forums on the Scratch website to find the answers you need.

Let's start creating your first Scratch program!

2

RAINBOW LINES . . .
IN SPACE!

n this chapter, you'll create a cool-looking
animation: a rainbow *V* that flies through
space and leaves colorful trails behind. This
program was inspired by the *demoscene*, a subculture
of elite programmers who made amazing graphical
programs starting in the 1980s.

Demosceners made beautiful, elaborate programs called *demos* that showed off their artistic, musical, and programming skills. But most amazing of all, these programs were *tiny*—just a few kilobytes! The program we'll write isn't quite as small, but it's dramatic and colorful, and it uses only a few lines of Scratch code.

Before you start coding, look at the following figure to see what the final program will look like. Then, go to *https://www .nostarch.com/scratch3playground/* to play the animation.

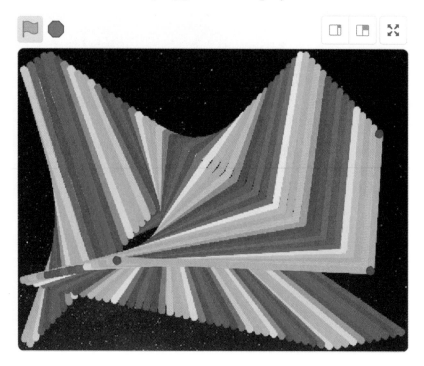

Just like demosceners, you can make beautiful programs. Let's create our own graphics demo in Scratch.

SKETCH OUT THE DESIGN

The first step of turning an idea into a Scratch program is to sketch out what you want it to look like. Planning your program helps you figure out the sprites you want and how they'll behave. I recommend drawing your ideas on paper so you can cross them out if you don't like them and write down notes and reminders. It's also best to keep the project simple.

When you've completed your simple game, you can then build on it to add complexity, which is the idea behind *iterative development*. First, you make the program work. Then, you make the program better. You can always add cool things to the basic program after you finish it.

Don't worry about making the sketch look nice. It's more important to have a solid plan for the main parts of the program. In my sketch, I have three parts: A, B, and C. We will work on these parts one at a time until we've built the full program.

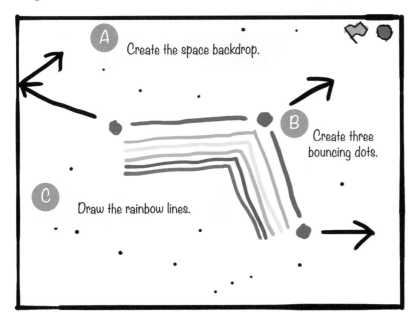

After you've completed a sketch of what you want your program to do, you can start programming. Go to *https:// scratch.mit.edu/*, sign up for an account on the site if you haven't already, and log in (having an account lets you save your programs on the Scratch website). After you've logged in, click the **Create** button at the top of your screen to start making your own Scratch project. Then, click the text field at the top to change the name of the project from *Untitled* to *Rainbow Lines*. Let's begin by tackling Part A of the sketch.

A CREATE THE SPACE BACKDROP

First, let's clean up sprites we won't use and set a background.

1. Clean Up and Set the Stage

Every time you create a new Scratch project, you'll see an orange cat sprite on a blank, white Stage. We don't need the cat sprite for our program, so right-click the Sprite1 cat in the Sprite List and either select **delete** to remove the cat from the Stage and the Sprite List or click the trash can icon next to the sprite.

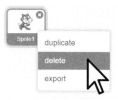

Click the **Choose a Backdrop** button (which looks like a landscape painting) under *New backdrop.*

The Backdrop Library window will open and display all the backdrops in alphabetical order. Click the **Stars** backdrop.

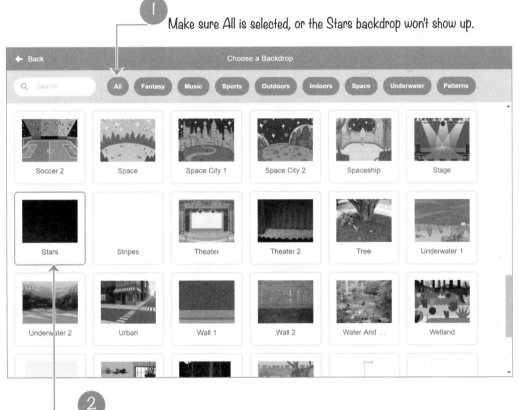

1 Make sure All is selected, or the Stars backdrop won't show up.

2 Select the Stars backdrop.

Now the Stage's backdrop looks like outer space.

B CREATE THREE BOUNCING DOTS

Next, we'll add three new sprites that represent the three points of the flying *V.*

2. Paint the Dot

Click the **Paint** button (which looks like a paintbrush).

A new sprite named Sprite1 is created in the Sprite List. Clicking this button also switches the editor to the Costumes tab, which contains the Paint Editor. Use the Brush tool to draw a small red dot near the crosshairs in the Paint Editor. It might help to zoom in by clicking the Zoom button (which looks like a magnifying glass) in the Paint Editor.

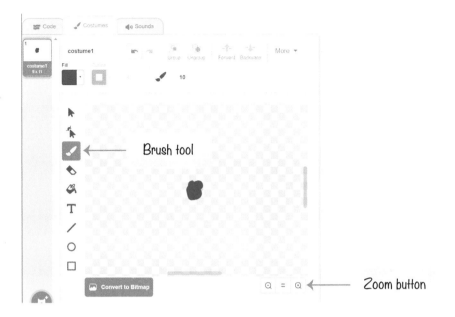

In the Sprite Pane above the Sprite List, change the sprite's name from Sprite1 to Dot 1.

Type the new name here.

Sprite   Dot 1   ↔ x   36   ↕ y   28

Show   ⊙ Ø   Size   100   Direction   90

Dot 1

3. Add Code for the Dot 1 Sprite

Now we can start programming. Click the **Scripts** tab to make the Code Area visible. Add the following code to the Code Area. You can find these blocks in the *Events* (yellow), *Motion* (dark blue), *Operators* (green), and *Control* (light orange) categories. If you have trouble understanding how to drag these blocks, view the animation at *https://www.nostarch.com/scratch3playground/*.

When you click the green flag, the Dot 1 sprite points in a random direction between −180 and 180 degrees. Then, the **forever** loop moves the sprite forward 10 steps and makes the sprite bounce when it hits the edge of the Stage. The sprite will continue to do this forever.

Notice that the Dot 1 sprite isn't drawing any rainbow lines yet. We'll do that later when we've created more sprites.

EXPLORE: DIRECTION AND DEGREES

Words like *up* or *right* are perfectly understood as directions by humans like you and me. But the computer needs a number to indicate an exact direction. All sprites in Scratch have their own direction number. The direction numbers are between −180 and 180 degrees. Pointing at 0 degrees is facing up. Pointing at 90 degrees is facing to the right. The following figure shows several directions and their degrees. Notice that the degrees increase in the clockwise direction and decrease in the counterclockwise direction. Also, notice that −180 and 180 degrees point in the same direction: down.

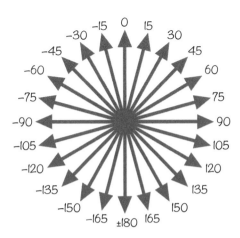

The **pick random -180 to 180** block chooses a random number to use as the direction. Then, the **point in direction** block points the sprite in that direction. This means that the sprite could be pointed in any possible direction.

Let's write a new script that demonstrates how degrees work. Open a new tab by pressing CTRL-T in your web browser and go to *https://scratch.mit.edu/* to open a new Scratch editor. You can edit multiple Scratch programs at the same time.

In the Code Area for the cat sprite named **Sprite1**, add the following code using blocks from the *Events* (yellow), *Control* (light orange), *Motion* (dark blue), and *Looks* (purple) categories. Keep in mind that we're writing a totally separate program from the *Rainbow Lines* program.

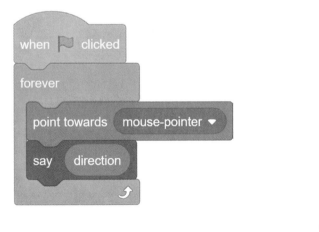

(continued)

When you run this program, the cat sprite points toward the mouse. The cat will say the direction in which it's pointing.

Notice that the direction number changes as the cat's direction changes.

4. Duplicate the Dot 1 Sprite

Right-click the Dot 1 sprite in the Sprite List and select **duplicate**. Do this twice so that you make two duplicates: Dot 2 and Dot 3. (Scratch automatically names the sprites with increasing numbers.)

SAVE POINT

Click the green flag to test the code so far. Check that all three dots are moving and bouncing around the Stage. When you duplicated the sprites, you also duplicated the sprite code. Click the red stop sign and save your program.

Ⓒ DRAW THE RAINBOW LINES

Now that we've created all the bouncing dots, we can create a fourth dot sprite to draw the rainbow lines. We'll write a program that makes this drawing dot move quickly between the three bouncing dot sprites, drawing a line as it moves. This process will repeat three times, and then after 10 seconds, the screen will clear.

5. Add the Code for the Drawing Dot Sprite

This project uses *Pen* code blocks, which are not included in the Block Palette by default. You must click the **Add Extension** button in the lower-left corner of the editor and select **Pen** from the window that appears.

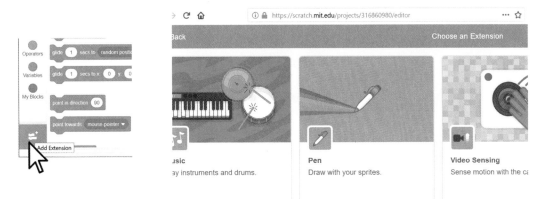

Right-click one of the bouncing dot sprites and select **duplicate**. Because this is a duplicate of the bouncing dots, it has some code we need to delete. In its Code Area, delete all the code blocks by dragging the **when green flag clicked** block (and all the blocks underneath it) to the Block Palette in the middle of the editor, where they'll disappear.

In the Sprite Pane, rename this sprite Drawing Dot.

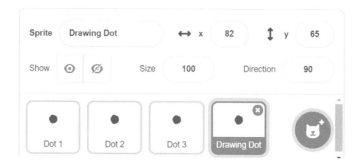

Add the following two scripts to the Drawing Dot sprite. You can find these blocks in the *Events* (yellow), *Pen* (clear), *Control* (light orange), and *Motion* (dark blue) categories. Remember, the *Pen* code blocks appear only if you clicked the Add Extension button and selected Pen.

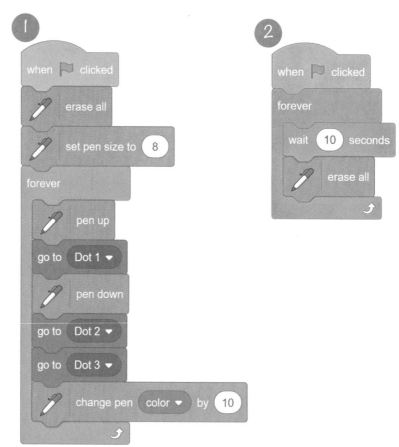

In script ❶, make sure you use the **go to** block, not the **go to x y** block. Also, be sure to change the **go to** blocks so they aren't going to the mouse pointer. To do so, click the white triangle on the block and select a sprite from the menu.

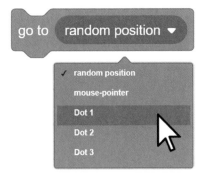

Before you run the code, let's talk through how it works. When you click the green flag in script ❶, the Drawing Dot sprite runs the **erase all** block to clear away any pen drawing already on the Stage. Then, the script runs the **pen down** block: as the sprite moves around, it draws a line on the Stage.

To better understand what the **pen down** block does, imagine holding down a giant marker on the floor while you walked around it: the marker would draw a line on the floor following you! The drawing dot goes to Dot 1, puts its pen down, goes to Dot 2, and then goes to Dot 3. Next, the **change pen color by 10** block changes the color of the pen slightly. (You can increase this number to make the colors change faster.) At the same time, the Dot 1, Dot 2, and Dot 3 sprites continue to move around on their own. So the *V* line that the Drawing Dot sprite draws also moves around.

Script ❷ is a lot simpler to understand. This code waits 10 seconds and then clears the screen of any marks made by the *Pen* blocks. Now the Stage won't become over-crowded with rainbow lines.

THE COMPLETE PROGRAM

The final code for the entire program is shown here. Notice that the code for the Dot 1, Dot 2, and Dot 3 sprites is identical. If your program isn't working correctly, check your code against this code:

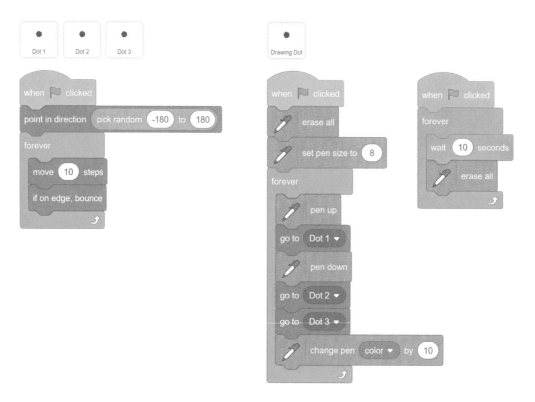

TURBO MODE

If you hold SHIFT while clicking the green flag, you can start the program in Turbo Mode. The computer is usually able to run code blocks quickly, but a program that draws sprites to the screen slows down the computer. In Turbo Mode, instead of drawing the screen after each code block, Scratch draws to the screen only after several code blocks. Human eyes won't notice the skipped drawings, and the program will look like it's running faster.

SHIFT-click the green flag to run the *Rainbow Lines* program in Turbo Mode. Almost instantly, the screen fills up. To end Turbo Mode, SHIFT-click the green flag again.

SUMMARY

In this chapter, you built a project that

▶ Has custom sprites you created (even if they are just dots)

▶ Uses the **pick random** block to point a sprite in a random direction

▶ Makes sprites move and bounce off the edges of the Stage

▶ Duplicates sprites and their code

▶ Uses the *Pen* blocks to draw rainbow lines

This project is a demo that users can watch but can't control. In Chapter 3, you'll make a maze game that lets players interact with the program by using the keyboard. This will be the first real game project in the book.

REVIEW QUESTIONS

Try to answer the following practice questions to test what you've learned. You probably won't know all the answers off the top of your head, but you can explore the Scratch editor to figure out the answers. (The answers are also online at *http://www.nostarch.com/scratch3playground/*.)

1. What happens when a sprite moves after it has run the **pen down** block?

2. If some code moves a sprite but no line is drawn behind it, what might cause this problem?

3. Which block causes the lines in the *Rainbow Lines* program to look like a rainbow?

4. Which code block do you use to make the rainbow lines thicker?

5. How do you turn on Turbo Mode? How do you turn it off?

6. How do you duplicate a sprite and its code blocks?

7. Where does a sprite point when its direction is 90 degrees?

8. What is the degree direction for pointing up?

9. You want a sprite to point down and move in that direction. In which color of blocks category would you find code blocks to do this?

10. How do you select a new backdrop from Scratch's Backdrop Library?

11. You see a sprite named Sprite1 in the Sprite List. How do you rename this sprite?

3

MAZE RUNNER

 ou've probably played a maze game before, but have you ever tried making one? Mazes can be tricky to complete, but they're easy to program. In this chapter, you'll create a game that lets the player guide a cat through a maze to reach its goal—a delicious apple! You'll learn how to move the cat with the keyboard and how to block its progress with walls.

Before you start coding, take a look at the final program. Go to *https://nostarch.com/scratch3playground/* and play the game.

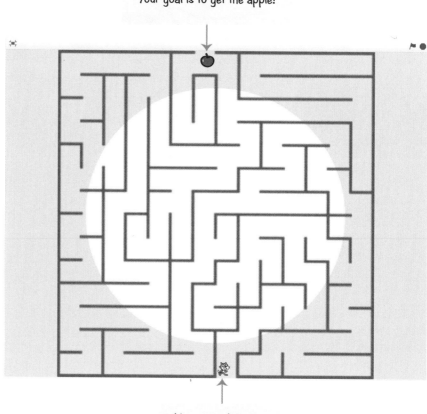

Your goal is to get the apple!

You control the cat.

SKETCH OUT THE DESIGN

First, draw what you want the game to look like on paper. With some planning, you can make your maze game a-maze-ing. (I never apologize for my puns.) My sketch for the maze game looks like the following figure.

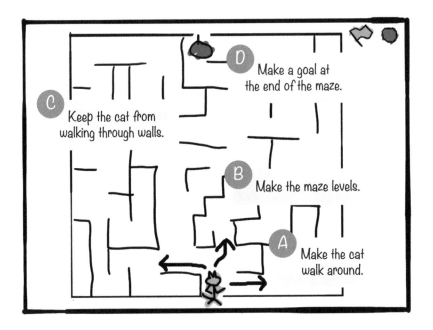

D Make a goal at the end of the maze.

C Keep the cat from walking through walls.

B Make the maze levels.

A Make the cat walk around.

If you want to save time, you can start from the skeleton project file, named *maze-skeleton.sb3*, in the resources ZIP file. This file has parts of the project already set up and only needs the code blocks added to it. Go to *https://nostarch.com/scratch3playground/* and download the ZIP file to your computer by right-clicking the link and selecting **Save link as** or **Save target as**. Extract all the files from the ZIP file. The skeleton project file has all the sprites already loaded, so you'll only need to drag the code blocks into each sprite. Click **File ▶ Load from your computer** in the Scratch editor to load the *maze-skeleton.sb3* file.

Even if you don't use the skeleton project, you should download the ZIP file from the website. This file contains the maze images you'll use in this chapter.

If you want to create everything on your own, click **File ▶ New** to start a new Scratch project. In the text field in the upper left, rename the project from *Untitled* to *Maze Runner*.

MAKE THE CAT WALK AROUND

In the *Maze Runner* game, the player will control the cat sprite. In Part A, you'll set up the code to control the cat with the arrow keys on the keyboard.

EXPLORE: X- AND Y-COORDINATES

To make the cat move around the Stage, you need to use coordinates. *Coordinates* are numbers that represent an exact location. The x-coordinate (also called *x position*) is a number that represents how far left or right a sprite is on the Stage. In other words, x is the sprite's *horizontal* position. The y-coordinate (also called *y position*) is a number that represents how far up or down a sprite is on the Stage. The y-coordinate is a sprite's *vertical* position.

Used together, x- and y-coordinates indicate a sprite's precise location on the Stage. The x-coordinate always comes first, and the coordinates are separated by a comma. For example, an x-coordinate of 42 and a y-coordinate of 100 would look like this: (42, 100).

In the center of the Stage is a point marked (0, 0), which is called the *origin*. In the following figure, I'm using the xy-grid backdrop from the Scratch Backdrop Library. (To load the xy-grid backdrop, click the **Choose a backdrop** button in the lower right and select the backdrop.) I've added several cat sprites who are all saying their x- and y-coordinates.

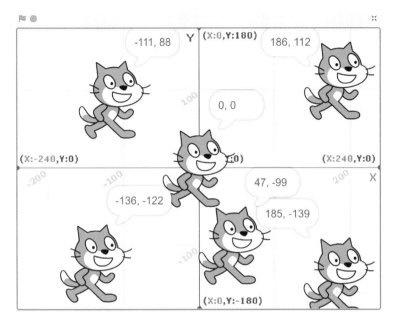

The rightmost side of the Stage has an x-coordinate of 240. The x-coordinates get smaller as you go left. In the center, the x-coordinate is 0. To the left of the center, the x-coordinates become negative numbers. The leftmost side of the Stage has an x-coordinate of −240. The y-coordinates work the same way: the top of the Stage has a y-coordinate of 180, the center is 0, and the bottom is −180.

Scratch displays the x- and y-coordinates of the currently selected sprite in the upper-right corner of the Sprite List. Sprites move around the Stage when you change their x- and y-coordinates, as shown here:

To make a sprite go . . .	change its . . .	by a . . .
Right	x-coordinate	positive number
Left	x-coordinate	negative number
Up	y-coordinate	positive number
Down	y-coordinate	negative number

(continued)

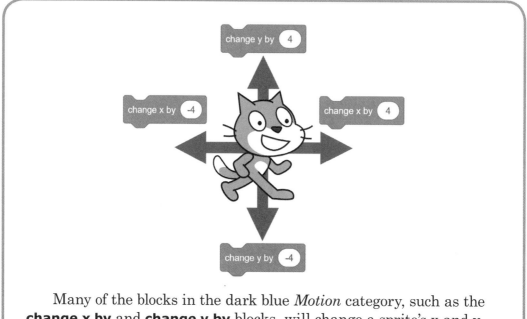

Many of the blocks in the dark blue *Motion* category, such as the **change x by** and **change y by** blocks, will change a sprite's x and y position. Note that changing a coordinate by a negative number is the same as subtracting a positive number from it.

I. Add Movement Code to the Player Sprite

The first bit of code you'll add will make the arrow keys move the cat sprite, which is named Sprite1. But first, click and rename this sprite Orange Cat. Then add the following code. You'll find these blocks in the *Events*, *Control*, *Sensing*, and *Motion* categories.

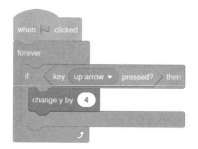

This code repeatedly checks whether the key is being pressed. The code literally reads "for forever, check if the up arrow key is pressed, and if it is, then change y by 4." If the up arrow key is not being pressed, Scratch skips the code inside the **if then** block.

Pressing the up arrow key makes the cat sprite move up. The **forever** loop block means Scratch will check over and over again whether the up arrow key is being pressed. This continues until you click the red stop sign.

The **forever** block is needed for this program to work. Without it, Scratch would check only *once* if the up arrow key was pressed. Then the program would end. But you want Scratch to keep checking if the up arrow key is pressed so the game doesn't end and the cat can get the apple. If your program doesn't seem to be doing anything, make sure you didn't forget to add the **forever** block.

When you code this on your own, be sure you use the **change y by** code block instead of the **change x by** or **set y to** code block. If your program isn't working correctly, check that your code is the same as the code in this book.

SAVE POINT

Click the green flag and try moving the cat by pressing the up arrow key. Then click the red stop sign and save your program.

2. Duplicate the Movement Code for the Cat Sprite

Now you'll add code for the other three arrow keys: down, left, and right. This code is similar to the code to move the cat sprite up. To save time, you can right-click or long press the **orange if then** block and select **Duplicate** to create a copy of the blocks. These blocks will be identical, so all you'll need to change are the dark blue *Motion* blocks for the other directions. Duplicating blocks can often be faster than dragging new ones from the Block Palette.

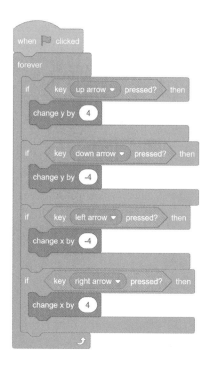

Scratch will now check if the four arrow keys are held down, one after another. After checking the right arrow key, Scratch starts back at the top of the loop and checks the up arrow key again. The computer checks them so fast that to human eyes, it looks like all of the arrow keys are being checked at the same time!

SAVE POINT

Click the green flag to test the code so far. The cat should walk up, down, left, and right when you press the arrow keys. Notice that Orange Cat is small enough to fit in the maze you'll create next. Click the red stop sign and save your program.

If your program doesn't work and you don't know how to fix it, you can start over by using the *maze-part-a.sb3* Scratch project file, which is in the resources ZIP file. Click **File ▶ Load from your computer** in the Scratch editor to load the *maze-part-a.sb3* file and then move on to Part B.

Ⓑ MAKE THE MAZE LEVELS

Next, we'll create the maze sprite and set the backdrop. The maze game would quickly get boring if it had only one maze, so we'll also add multiple levels to the game.

3. Download the Maze Images

You could draw the maze sprite yourself, but let's use images from the ZIP file instead. One of the maze images is the *Maze .sprite3* file.

In the Scratch editor, click the **Upload Sprite** button, which appears after you tap or hover over the **Choose a Sprite** button (which looks like a face) and select *Maze.sprite3* to upload the file. This creates a new sprite named Maze with several maze costumes. Every sprite in Scratch can have multiple costumes to change the way it looks. These costumes, which you can see by clicking the **Costumes** tab, are often used to animate the sprite.

Your Sprite List should look like this:

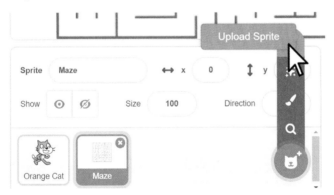

4. Change the Backdrop

Let's add a little flair to the maze by placing some artwork in the background. You can use whichever backdrop you like. Change the Stage's backdrop by clicking the **Choose a Backdrop** button in the lower right to open the Scratch Backdrop Library window. Choose a backdrop (I chose Light).

5. Start at the First Maze

Add the following code to the Maze sprite. You can find these blocks in the *Events*, *Looks*, and *Motion* categories.

Each of the Maze sprite's costumes will be a new level. When the player clicks the green flag to start the program, the game should begin with the first costume and make sure the maze is in the center of the Stage. We will add code to switch to the next level in steps 8 and 9.

Note that the Code Area shows code blocks for the selected sprite. Be sure the Maze sprite is selected in the Sprite List; otherwise, you'll add the maze's code to a different sprite. Each sprite needs its own code to work correctly. If you don't see maze1 in the **switch costume to** block, the Orange Cat sprite is most likely selected.

If your Scratch program doesn't work and you don't know how to fix it, you can start over by using the *maze-part-b.sb3* Scratch project file, which is in the resources ZIP file. Click **File ▶ Load from your computer** in the Scratch editor to load the *maze-part-b.sb3* file and then move on to Part C.

Ⓒ KEEP THE CAT FROM WALKING THROUGH WALLS

When you click the green flag now, you'll be able to move the cat through the maze. But you'll also be able to move the cat through the walls of the maze, because nothing in the program prevents this from happening. The code only states, "When the right arrow key is pressed, move the cat right." The cat moves, whether or not a wall is there.

6. Check Whether the Cat Is Touching the Walls

Let's add code that checks whether the cat is touching a blue wall. If it is, the cat should back away. So if the cat moves to the right and is touching a wall, it should automatically move left. This will undo the player's move and prevent the cat from moving through the wall. Click the Orange Cat sprite in the Sprite List and modify the code to look like the following. Notice that we're using the **touching?** block, not the **touching color?** block.

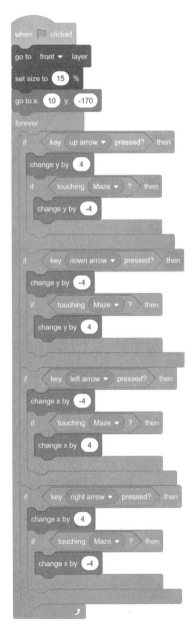

Also, you might have noticed that the Orange Cat sprite is Godzilla sized compared to the maze, making the cat look unrealistic. Add a **set size** block from the *Looks* category to make the Orange Cat sprite smaller. You also want the Orange Cat sprite to always be shown on top of the maze, so you'll add the **go to front layer** block. Put these two code blocks at the top of the script.

SAVE POINT

Click the green flag to test the code so far. Make sure the Orange Cat sprite cannot walk through the maze walls. Test this for all four directions. Then click the red stop sign and save your program.

If your Scratch program doesn't work and you don't know how to fix it, you can start over by using the *maze-part-c.sb3* Scratch project file, which is in the resources ZIP file. Click **File ▶ Load from your computer** in the Scratch editor to load the *maze-part-c.sb3* file and then move on to Part D.

Ⓓ MAKE A GOAL AT THE END OF THE MAZE

Right now, it isn't clear where the player is supposed to end up in the maze. Let's add an apple at the other end of the maze to make the player's goal more obvious.

7. Create the Apple Sprite

Click the **Choose a Sprite** button. When the Sprite Library window appears, select **Apple** to add a new sprite named Apple to the Sprite List.

When the game starts, you want the Apple sprite to move to the end of the maze at the top of the Stage. The Apple sprite also has to be small enough to fit in the maze. Add the following code to the Apple sprite:

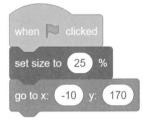

8. Detect When the Player Reaches the Apple

The *Maze Runner* game has a sprite for the player, the maze, and the apple at the end of the maze. Now it just needs the code to detect when the player reaches the end. When that happens, you'll have Scratch play a sound and then swap costumes to the next level. But before you add the code, you need to load the Cheer sound. Select Orange Cat in the Sprite List. Click the **Sounds** tab at the top of the Block Palette and then click the **Choose a Sound** button in the lower left. This button looks like a speaker.

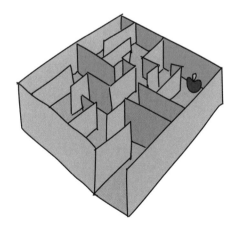

In the Sound Library window that appears, select Cheer to load the Cheer sound. Now click the **Code** tab.

The **broadcast** block causes a script under a matching **when I receive** block to run. Add this script to the Orange Cat sprite's code:

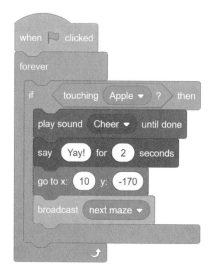

To make the **broadcast** block broadcast the next maze message, click the white triangle in the **broadcast** block and select **new message**.

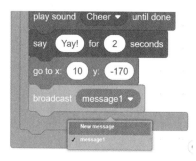

In the window that appears, type *next maze* as the message name and click **OK**.

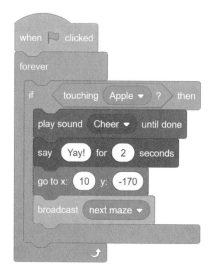

9. Add the Broadcast Handling Code to the Maze Sprite

Click the Maze sprite in the Sprite List and add the following code to it:

This code changes the maze level when it receives the next maze broadcast.

THE COMPLETE PROGRAM

The following code shows the entire program. If your program isn't working right, check your code against this complete code. The complete program is also in the resources ZIP file as the *maze.sb3* file.

I hope the code for this maze program isn't too labyrinthine.

Orange Cat

```
when [flag] clicked
go to [front v] layer
set size to (15) %
go to x: (10) y: (-170)
forever
    if < key (up arrow v) pressed? > then
        change y by (4)
        if < touching (Maze v) ? > then
            change y by (-4)
    if < key (down arrow v) pressed? > then
        change y by (-4)
        if < touching (Maze v) ? > then
            change y by (4)
    if < key (left arrow v) pressed? > then
        change x by (-4)
        if < touching (Maze v) ? > then
            change x by (4)
    if < key (right arrow v) pressed? > then
        change x by (4)
        if < touching (Maze v) ? > then
            change x by (-4)
```

```
when [flag] clicked
forever
    if < touching (Apple v) ? > then
        play sound (Cheer v) until done
        say (Yay!) for (2) seconds
        go to x: (10) y: (-170)
        broadcast (next maze v)
```

Maze

```
when [flag] clicked
switch costume to (maze1 v)
go to x: (0) y: (0)
```

```
when I receive (next maze v)
next costume
```

Apple

```
when [flag] clicked
set size to (25) %
go to x: (-10) y: (170)
```

VERSION 2.0: TWO-PLAYER MODE

Now that the basic maze game is working, you can do some iterative development and add small improvements one at a time. Iterative development helps you avoid making a game that is too large for you to finish.

In version 2.0 of *Maze Runner*, you'll add a second player. The two players will race against one another. The first player starts at the bottom and races to the top; the second player races from the top to the bottom. Because they both must travel the same path, the distance for each is the same.

Duplicate the Apple Sprite

The second player needs a goal too. Right-click or long press the Apple sprite and select **Duplicate** to make a copy of the Apple sprite and its code. The new sprite is automatically named Apple2. Select the Apple2 sprite and click the **Costumes** tab. Select a green fill color and then select the **Fill** tool to the right (it looks like a tipped cup). Then click the red part of the apple to change it to green. When you're done, Apple2 will look like the following figure.

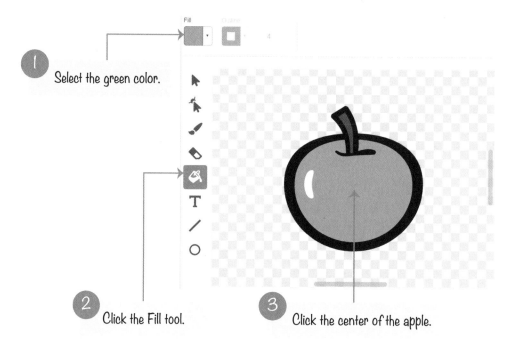

1. Select the green color.

2. Click the Fill tool.

3. Click the center of the apple.

Modify the Apple2 Sprite's Code

You need to modify the Apple2 sprite's code to look like the following code so that the green apple starts at the bottom of the maze rather than the top:

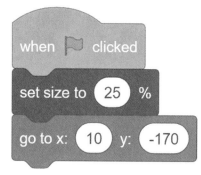

Duplicate the Orange Cat Sprite

Now let's add a second cat sprite. Right-click or long press the Orange Cat sprite and select **Duplicate** from the menu to make a copy of Orange Cat and its code. The new sprite is automatically named Orange Cat2. The Orange Cat and Orange Cat2 sprites need to look different enough that the players can tell them apart. Similar to what you did for Apple2, click the **Costumes** tab and change Orange Cat2 from orange to blue.

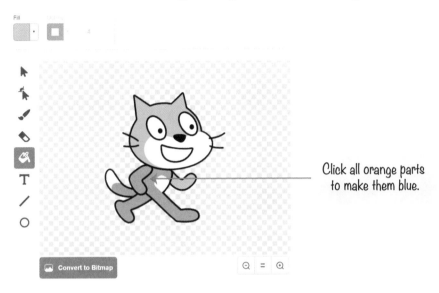

Click all orange parts to make them blue.

In the Sprite Pane, rename Orange Cat2 to Blue Cat.

Modify the Code for the Blue Cat Sprite

Right now the Blue Cat sprite has the same code as the Orange Cat sprite. You'll need to change the code; otherwise, the arrow keys will control both player 1 and player 2. The second player will control the Blue Cat sprite using the WASD keys (pronounced *whaz-dee*). The W, A, S, and D keys are often used as a left-handed version of the up, left, down, and right arrow keys.

Change the two **go to x y** blocks and the **key pressed?** blocks for Blue Cat to look like the following code. Also, remember to change **if touching Apple** to **if touching Apple2**.

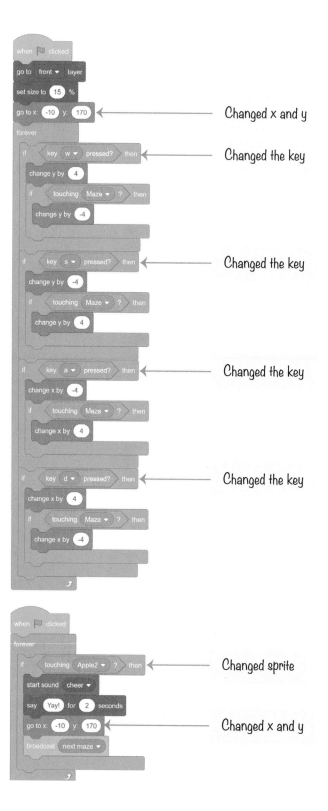

Changed x and y

Changed the key

Changed the key

Changed the key

Changed the key

Changed sprite

Changed x and y

Go Back to the Starting Position

The sprites go back to their starting positions when they touch their apples, but they also need to go there when the other cat wins. Add the following script to the Orange Cat sprite:

Then add the following script to the Blue Cat sprite:

This way, when the other cat wins and broadcasts the next maze message, both cats will go back to their starting positions.

SAVE POINT

Click the green flag to test the code so far. Try moving both players using the arrow keys and the WASD keys. Make sure each of the eight keys moves the correct cat and only that cat. Try playing the full game. Make sure the level changes to the next maze when the second player touches the green apple. Make sure both players are sent to their starting points when the next level begins. Click the red stop sign and save your program.

You've just upgraded your maze game to support two players. Find a friend to race against. Player 1 uses the arrow keys, and player 2 uses the WASD keys.

CHEAT MODE: WALK THROUGH WALLS

Teleporting is a cool cheat, but players can't control where they teleport to. Also, it would be too obvious that a player was cheating if they suddenly moved across the Stage and through many walls. However, you can add a subtler cheat that lets the cats move through walls when a special key is held down.

Add the Walk-Through-Walls Code to Orange Cat

For the Orange Cat sprite, modify the walking code so the **touching Maze?** blocks are replaced with the **touching Maze?** and **not key l pressed?** blocks. Only the code for the up arrow is shown here, but you'll want to replace the **touching Maze?** blocks in all four of the **if key pressed** cases.

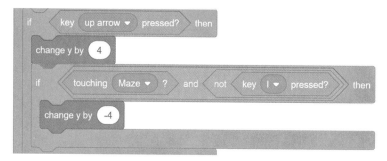

This enables the wall-blocking code only if the L key is *not* being pressed. If the L key *is* pressed, this wall-blocking code is skipped, and the player walks through the walls.

Add the Walk-Through-Walls Code to Blue Cat

Make the same walk-through-walls code changes to the Blue Cat sprite, except instead of **key l pressed**, make it **key q pressed**. The second player can walk through walls when the Q key is held down.

SAVE POINT

Click the green flag to test the code so far. Try walking through the walls by holding down the L or Q key. Make sure both cats can walk through walls but only when holding down the correct key. Click the red stop sign and save your program.

With this cheat code in the program, you can hack the game and make your cat walk through walls. This cheat shows you that anything is possible in your Scratch programs.

SUMMARY

In this chapter, you built a game that

▶ Has a cat sprite that can walk up, down, left, and right when the player presses certain keys

▶ Has walls that the sprites can't walk through

▶ Broadcasts messages from one sprite that another sprite can receive

▶ Has a maze sprite with eight different costumes

▶ Supports two players using different keyboard keys

▶ Includes cheat modes that let the cats walk through walls

A two-player game is more exciting than a single-player game. Now, instead of just solving a maze, you are racing against another player! And you get to show off your Scratch game to someone else.

In Chapter 4, you'll work with a basketball game. This game uses a side view, unlike the maze's bird's-eye view. But this means you'll be able to add jumping and gravity, which are great techniques to use in many types of Scratch games.

REVIEW QUESTIONS

Try to answer the following practice questions to test what you've learned. You probably won't know all the answers off the top of your head, but you can explore the Scratch editor to figure out the answers. (The answers are also online at *https://nostarch.com/scratch3playground/*.)

1. Which block will change the size of a sprite?

2. How can the code in one sprite send a message to another sprite to do something?

3. How might you use the WASD keys on the keyboard?

4. How can you duplicate some code blocks from one sprite to another sprite?

5. What will happen if you accidentally use a **change y by** code block instead of a **change x by** code block?

6. If you want a sprite to play the Cheer sound, how do you load this sound?

7. Look at the following code. It lets the player press the arrow keys to move the sprite left and right. It works, but what would you change to make the sprite walk faster?

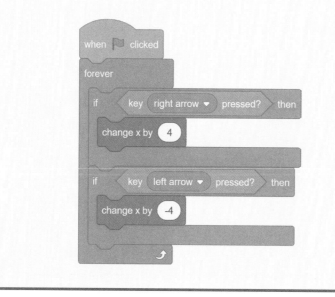

4

SHOOTING HOOPS
WITH GRAVITY

any platformer games, such as *Super Mario Bros.* and *Sonic the Hedgehog*, show the action from the side, with the ground at the bottom of the screen and the character appearing in profile. These games have *gravity*: characters can jump up and then drop until they land on the ground. In this chapter, we'll write a basketball game that has gravity. The player will jump and throw a basketball, and the basketball player and ball will drop to the ground.

Before you start coding, look at the final program at *https://nostarch.com/scratch3playground/*.

SKETCH OUT THE DESIGN

First, let's sketch out what we want the game to do. The player controls the cat, which can move left and right and jump. The goal is to have the cat shoot baskets into a moving basketball hoop, which will glide around the Stage randomly.

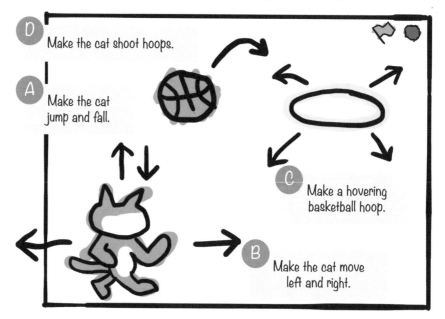

If you want to save time, you can start from the skeleton project file, named *basketball-skeleton.sb3*, in the resources ZIP file. Go to *https://nostarch.com/scratch3playground/* and download the ZIP file to your computer by right-clicking the link and selecting **Save link as** or **Save target as**. Extract all the files from the ZIP file. The skeleton project file has all the sprites already loaded, so you'll only need to drag the code blocks into each sprite.

Ⓐ MAKE THE CAT JUMP AND FALL

Let's start by adding gravity to make the cat jump up and come back down.

I. Add the Gravity Code to the Cat Sprite

Rename the Sprite1 sprite to Cat in the Sprite List. Then rename the program from *Untitled* to *Basketball* in the text field at the top of the Scratch editor.

Click the **Choose a Backdrop** button in the lower right to open the Backdrop Library window. Select **Wall 1** to change the backdrop. The Stage will look like this:

Programming gravity requires *variables*. You can think of a variable as a box for storing numbers or text that you can use later in your program. We'll create a variable that contains a number representing how fast the cat is falling.

First, make sure the Cat sprite is selected in the Sprite List; then click the Scripts tab. In the orange *Variables* category, click the **Make a Variable** button to bring up the New Variable window. Enter y velocity as the variable name. (*Velocity* means how fast something is going and in what direction. When y velocity is a positive number, the cat moves up. When y velocity is a negative number, the cat moves down.) Make sure **For this sprite only** is selected. (If you see only **For all sprites**, the Stage is selected, not the Cat sprite.) Then click **OK**.

New Variable	
New variable name:	
y velocity	
◯ For all sprites	◉ For this sprite only ← Make sure For this sprite only is selected!
☐ Cloud variable (stored on server)	
	Cancel **OK**

Several new blocks will appear in the *Variables* category, and one of those is the round y velocity variable block.

EXPLORE: FOR ALL SPRITES VS. FOR THIS SPRITE ONLY

When you create the y velocity variable, you have to select **For this sprite only**. This option creates a variable that *only* the Cat sprite can use. The **For all sprites** option will create a variable that all sprites can use.

To identify the kind of variable you've created, check the box next to the round variable block in the Block Palette to make the variable appear on the Stage. If you selected **For this sprite only**, the sprite name will display in front of the variable name. But if you selected **For all sprites**, only the variable name will show.

Check this box to make the variable visible on the Stage.

The Cat sprite name before y velocity means that this is a For this sprite only variable.

(continued)

If you made a mistake and the sprite name Cat does not appear at the start of the variable name in your program, right-click the **y velocity** block in the orange *Variables* category and select **Delete the "y velocity" variable** from the menu. Then create the y velocity variable again, making sure you select **For this sprite only**.

As with any variable block, you can place the y velocity block anywhere you would normally enter a number or some text. The code block will use the number or text set in the variable. When you use variables, your program can change the variable's number or text *while the program is running*.

To put a value in your variable, you use the orange **set to** block. For example, if you created a variable greeting, you could use **set to** to put the value Hello! in it. Then, you could use greeting in a **say** block, which would be the same as entering Hello!. (Don't add these blocks or create a greeting variable in your *Basketball* program; this is just an example.)

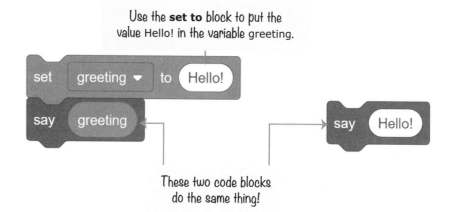

Use the **set to** block to put the value Hello! in the variable greeting.

These two code blocks do the same thing!

Using the variable lets the **say** block change its text while the program runs. If you want to change the greeting text while the program is running, you can add another **set to** block to your program. If the variable contains a number, you can add to or subtract from this number by using the **change by** block.

Gravity makes objects *accelerate* downward. In the game, the Cat sprite has to move down, and the speed at which it moves down must change *while the program is running*. Add the following code to the Cat sprite to add gravity to your Scratch program. This is the minimal amount of code you need to make a sprite fall under gravity. You could add this code to any sprite to make it fall.

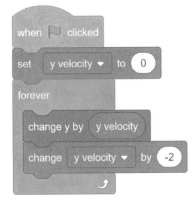

When you click the green flag, the y velocity variable is set to 0, and the script enters a **forever** loop. The y position (vertical position) of the Cat sprite is changed by y velocity, and y velocity is changed by -2. As the program goes through the loop, the y position will change faster and faster, making the cat fall faster and faster.

SAVE POINT

Before you click the green flag, drag the Cat sprite to the top of the Stage. When you click the green flag, notice that the cat begins to fall. If you want the cat to fall again, click the red stop sign, move the Cat sprite back to the top of the Stage, and click the green flag again. Save your program.

2. Add the Ground-Level Code

Right now, the cat falls. But we want the cat to stop when it hits the ground. Let's keep adding to the Cat sprite's code so it looks this:

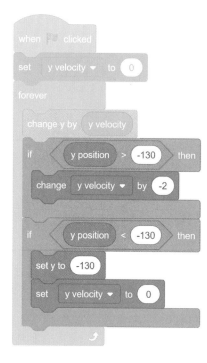

In this code, we set the **y position** of the ground level to -130. If the Cat sprite's y position is greater than (above) the ground level, then y velocity is changed by -2, and the Cat sprite will fall. Eventually, the Cat sprite will fall past -130, and its y position will be less than (below) the -130 ground level. When that happens, the Cat sprite will be reset at the ground level of -130, and y velocity will go back to 0 to stop the sprite from falling.

Click the green flag to test the code so far. Drag the cat up using your mouse and let go. Make sure the cat falls to the ground but does not fall past the edge of the Stage. You can experiment with different ground levels by changing -130 to another number. Then click the red stop sign and save your program.

When you're done testing your program, make the y velocity variable invisible on the Stage by unchecking the checkbox next to it in the orange *Data* category.

3. Add the Jumping Code to the Cat Sprite

After adding the gravity code to the Cat sprite, making the cat jump is easy. Add this code to the Cat sprite:

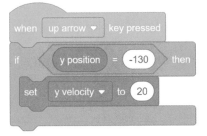

Now when you press the up arrow key, y velocity is set to the positive number 20, making the Cat sprite jump up. But the y velocity variable will still be changed by -2 each time the loop runs. So although the cat jumps by 20 at first, the next time through the loop, it will be at 18, then 16, and so on. Notice that the **if then** block checks that the Cat sprite is on the ground. You shouldn't be able to make the cat jump if it's already in midair!

When y velocity is set to 0, the Cat sprite is at the peak of the jump. Then y velocity changes by -2 each time through the loop, and the Cat sprite continues falling until it hits the ground. Try experimenting with different numbers for the **set y velocity to** and **change y velocity by** blocks. Figure out how to make the cat jump higher or lower (but always above the ground) or make gravity stronger or weaker.

Click the green flag to test the code so far. Press the up arrow key and make sure the cat jumps up and drops back down. Then click the red stop sign and save your program.

B MAKE THE CAT MOVE LEFT AND RIGHT

Let's add the Cat sprite's walking code next so that the player can control the cat with the keyboard.

4. Add the Walking Code to the Cat Sprite

Add the following code to the bottom of the Cat sprite's code:

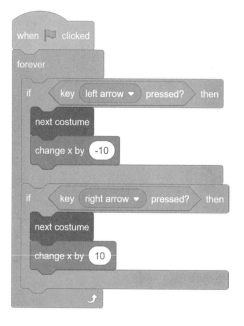

Inside the **forever** loop, the program checks whether the left arrow or right arrow key is being pressed. If one of these arrows is pressed, the Cat sprite switches to the next costume and changes its x position by -10 (moves to the left) or 10 (moves to the right). The Cat sprite comes with two costumes that you can view by clicking the Costumes tab above the Block Palette. By switching between the two costumes using the **next costume** block, you can make it look like the cat is walking.

SAVE POINT

Click the green flag to test the code so far. Press the left and right arrow keys, making sure the cat moves in the correct direction. It's okay if the cat walks backward when moving to the left—that's what we want. Then click the red stop sign and save your program.

C MAKE A HOVERING BASKETBALL HOOP

Now that the Cat sprite is complete, let's move on to the next sprite in the game: the basketball hoop.

5. Create the Hoop Sprite

Click the **Paint** button, which appears after you tap or hover over the **Choose a Sprite** button. The drawing tool buttons will appear on the left side of the Paint Editor. Select a yellow color and use the Circle tool to draw a hoop. You can also change the width in the text field next to the Fill and Outline color boxes to make the circle's line thicker. Make sure the Paint Editor's crosshairs are in the center of the hoop.

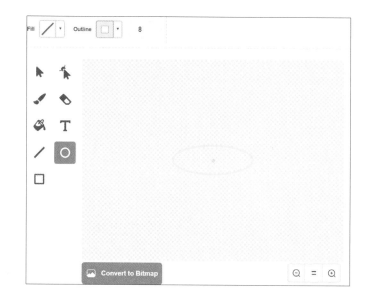

Rename the sprite Hoop in its Sprite Pane. We want the Hoop sprite to play a cheer sound when the player makes a basket, so let's load the Cheer sound next. Click the **Sounds** tab at the top of the Block Palette and then click the **Choose a Sound** button in the lower left. When the Sound Library window opens, select Cheer.

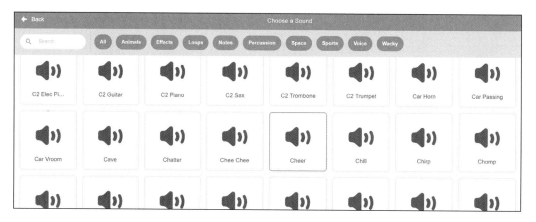

The Cheer sound will now appear as an option for the **start sound** block you'll add to the Hoop sprite.

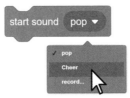

Add the following code to the Hoop sprite to make it glide randomly around the top half of the Stage. You'll need to create a broadcast message by clicking the **when I receive** block's white triangle and selecting **new message**. Name the new broadcast message swoosh.

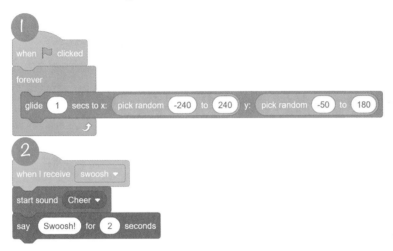

Script ❶ makes the hoop slide to a new position every second. A moving hoop will make the game more challenging to play! Script ❷ plays the Cheer sound and displays "Swoosh!" when the swoosh broadcast is received. Later, we'll make the Basketball sprite broadcast this message when a basket is made.

6. Create the Hitbox Sprite

Now let's think about how to create code that determines whether the player makes a basket. We could write a program that checks whether the basketball is simply *touching* the hoop. But if we did that, just the edges of the basketball touching the hoop would count as a basket. We want the basket to count only if the basketball goes through the middle of the hoop. We'll have to think of a better fix.

If we simply checked whether the basketball was touching the hoop, this would count as a basket. But that's not how basketball works!

Instead, you can create a *hitbox*. A hitbox is a game-design term for a rectangular area that determines whether two game objects have collided with each other. We'll make a hitbox sprite. Create a new sprite by clicking the **Paint** button, which appears after you tap or hover over the **Choose a Sprite** button. Draw a small black square in the middle of the crosshairs by using the Rectangle tool and selecting the solid fill option. Rename this sprite Hitbox. The Hitbox sprite will look like this:

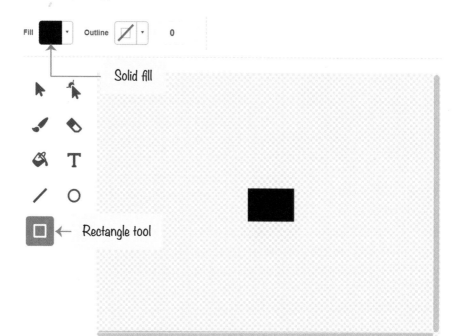

Add the following code to the Hitbox sprite:

The Hitbox sprite will now follow the Hoop sprite, no matter where it glides.

In step 9, we'll write a program that makes sure a basket counts only if the basketball is touching the Hitbox sprite, not the Hoop sprite. The basketball will have to be much closer to the middle of the hoop to count as a basket!

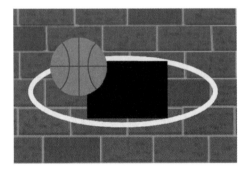

It looks odd to see a black square in the middle of the Hoop sprite, so let's make the Hitbox sprite invisible. Add the **set ghost effect** block (it originally appears in the Block Palette as **set color effect**, but you can change "color" to "ghost") to the Hitbox sprite and set it to 100.

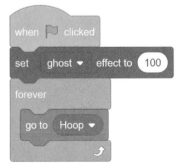

There's a difference between the **hide** block and the **set ghost effect to 100** block. If you used the **hide** block to make the Hitbox sprite invisible, the touching blocks would never detect that the ball was touching the Hitbox sprite, and the player would never be able to score. The **set ghost effect to 100** block makes the sprite invisible to you, but still allows the touching blocks to detect the Hitbox sprite's presence.

SAVE POINT

Click the green flag to test the code so far. Make sure the hoop glides around the Stage and the hitbox rectangle is always in the middle of it. Then click the red stop sign and save your program.

Ⓓ MAKE THE CAT SHOOT HOOPS

Next, you'll add a basketball for the cat to throw. Like the cat, the basketball will have gravity code and drop to the ground.

7. Create the Basketball Sprite

Click the **Choose a Sprite** button in the lower right to open the Sprite Library window. Select the **Basketball** sprite.

Next, click the **Sounds** tab at the top of the Block Palette. Then click the **Choose a Sound** button at the lower left to open the Sound Library window. Select the Pop sound. Click the **Scripts** tab at the top of the Block Palette to bring back the Code Area.

Now go to the orange *Variables* category. You'll make two variables. Click the **Make a Variable** button. Name the variable y velocity and make sure **For this sprite only** is selected before you click **OK**. Because they are **For this sprite only** variables, the Basketball sprite's y velocity variable is separate from the Cat sprite's y velocity variable. Even though they have the same name, they are two different variables.

Click the **Make a Variable** button again to make another variable named Player 1 Score, but this time select **For all sprites**. (We capitalize Player 1 Score because we'll make it visible on the Stage. Uncheck the box next to y velocity to hide it on the Stage.) The new variable blocks should appear in the orange *Variables* category.

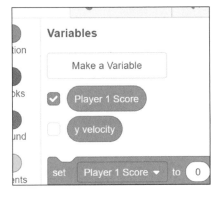

8. Add the Code for the Basketball Sprite

After you've added the pop sound and two variables, add the following code to the Basketball sprite.

Script ❶ makes sure the player starts with 0 points and hides the Basketball sprite to start.

Script ❷ uses code similar to the Cat sprite code. When the player presses the spacebar, the basketball appears in front of the cat and starts moving forward. The code sets the Basketball sprite's y velocity variable to a positive number, just like how the Cat sprite's y velocity variable is set to a positive number when the cat jumps; this is how the cat throws the ball.

The **repeat until y position < -130** block will keep the Basketball sprite falling until it reaches the ground. When it reaches the ground, the basketball will be hidden until the next time the player presses the spacebar.

SAVE POINT

Click the green flag to test the code so far. Press the spacebar to make the cat throw the basketball. Make sure the basketball disappears when it touches the ground. Then click the red stop sign and save your program.

9. Detect Whether a Basket Is Made

Next, you'll add the code that checks whether the Basketball sprite is touching the Hitbox sprite. This is how you know whether a basket has been made, and if it has, the Player 1 Score variable should increase. However, there is one snag: the basket shouldn't count if the basketball goes *up* through the hoop.

Keep in mind that if the y velocity variable is positive, the **change y by y velocity** block will move the Basketball sprite up. If y velocity is 0, then the Basketball sprite is not moving up or down. But if y velocity is a negative number, then the Basketball sprite is falling.

So, you'll add another **if then** condition to the Basketball sprite's code. You'll increase the score only if the basketball is touching the hitbox (using the **touching Hitbox?** block) and moving *downward* (**y velocity < 0**).

```
when  space ▼  key pressed

start sound  pop ▼

go to  Cat ▼

set  y velocity ▼  to  24

show

repeat until  ( y position < -130 )

    change x by  8

    change y by  10

    change  y velocity ▼  by  -2

    turn ↻  6  degrees

    if  ( touching  Hitbox ▼  ? ) and ( y velocity < 0 )  then

        change  Player 1 Score ▼  by  1

        broadcast  swoosh ▼

hide
```

The **and** block combines two conditions. For Scratch to run
the code blocks inside the **if then** block, *both* of these conditions must be true. It isn't enough for the sprite to be touching
the Hitbox sprite *or* for the y velocity variable to be less than 0.
Both **touching Hitbox?** *and* **y velocity < 0** must be true for
the player to score. If these conditions are true, the Player 1
Score variable is increased by 1, and the swoosh message is
broadcast.

Click the green flag to test the code so far. Shoot some baskets. The Player 1 Score variable should increase only if the basketball touches the center of the hoop and is moving down. The Hoop sprite should also say *swoosh* and play the Cheer sound when this happens. Click the red stop sign and save your program.

10. Fix the Scoring Bug

Did you notice that Player 1 Score increases by several points for a single basket? This is a *bug*, which is a problem that makes the program behave in an unexpected way. We'll need to take a careful look at the code to figure out why this happens.

    The **repeat until** loop block keeps looping until the ball hits the ground, so all of this code is for a single throw. The **repeat until** loop block checks several times whether the Basketball sprite is touching the Hitbox sprite and falling. The Player 1 Score value should increase only the first time.

    You fix this bug by creating a new variable that keeps track of the first time the basketball touches the hoop for a single throw. Then you can make sure the player scores a point only once per throw.

    Click the orange *Variables* category in the Block Palette and then click **Make a Variable**. Name the variable made basket and select **For this sprite only**. Then modify the Basketball sprite code.

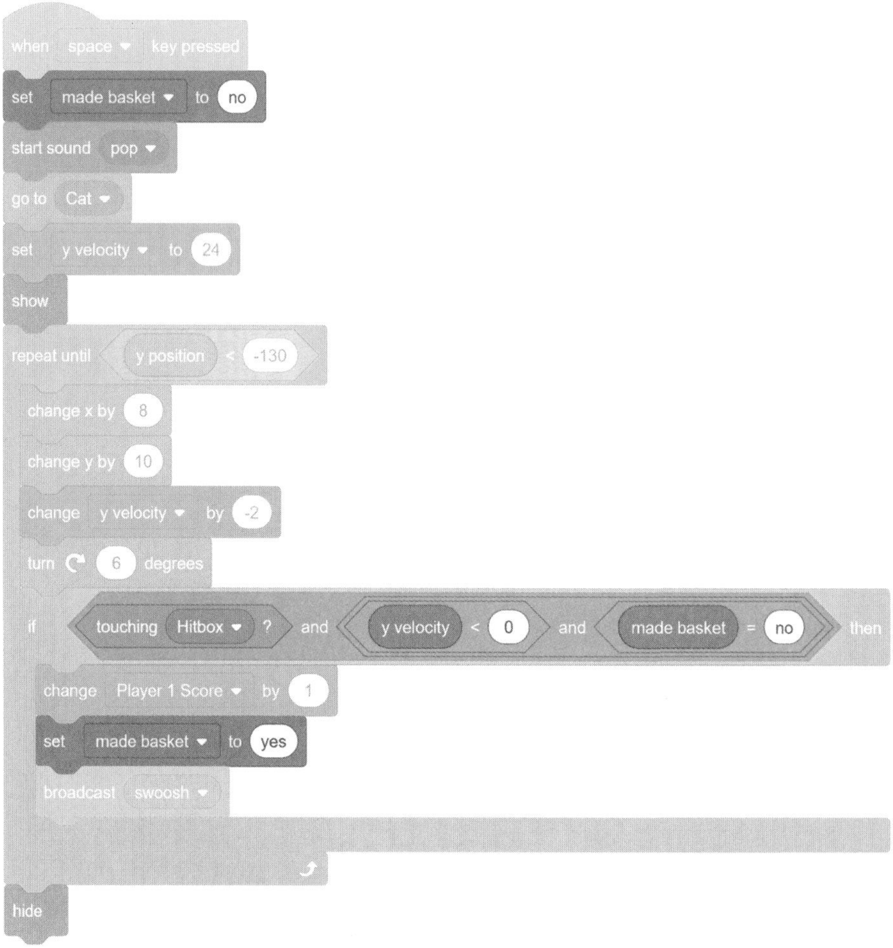

```
when  space ▼  key pressed

set   made basket ▼   to  no

start sound  pop ▼

go to  Cat ▼

set   y velocity ▼   to  24

show

repeat until  ( y position  <  -130 )

    change x by  8

    change y by  10

    change  y velocity ▼   by  -2

    turn ↻  6  degrees

    if  < touching  Hitbox ▼  ?  and  ( y velocity  <  0 )  and  ( made basket  =  no ) >  then

        change  Player 1 Score ▼   by  1

        set   made basket ▼   to  yes

        broadcast  swoosh ▼

hide
```

The made basket variable is set to no when the player first presses the spacebar. This makes sense, because the player has not made a basket when the ball is initially thrown. We'll also use an **and** block to add another condition to the code that checks whether a basket has been made. A basket is now detected, and the code in the **if then** block is run when three conditions are true:

- ▶ The Basketball sprite is touching the Hitbox sprite.
- ▶ The y velocity variable is negative (the basketball is falling).
- ▶ The made basket variable is set to no.

The first time the Basketball sprite detects it has made a basket, it increases Player 1 Score by 1 point and sets made basket to yes. In future checks for that shot, made basket will not be equal to no, so the basket is no longer detected. The made basket variable resets to no the next time the player presses the spacebar to throw the basketball.

SAVE POINT

Click the green flag to test the code so far. Shoot some baskets. Make sure that Player 1 Score increases by only one point for each basket. Then click the red stop sign and save your program.

THE COMPLETE PROGRAM

Here is the final code. If your program isn't working correctly, check your code against this code.

Cat

```
when [flag] clicked
set [y velocity ▼] to (0)
forever
  change y by (y velocity)
  if < (y position) > (-130) > then
    change [y velocity ▼] by (-2)
  if < (y position) < (-130) > then
    set y to (-130)
    set [y velocity ▼] to (0)
```

```
when [up arrow ▼] key pressed
if < (y position) = (-130) > then
  set [y velocity ▼] to (20)
```

```
when [flag] clicked
forever
  if < key (left arrow ▼) pressed? > then
    next costume
    change x by (-10)
  if < key (right arrow ▼) pressed? > then
    next costume
    change x by (10)
```

Hitbox

```
when [flag] clicked
set [ghost ▼] effect to (100)
forever
  go to (Hoop ▼)
```

Hoop

```
when [flag] clicked
forever
  glide (1) secs to x: (pick random (-240) to (240)) y: (pick random (-50) to (180))
```

```
when I receive (swoosh ▼)
start sound (Cheer ▼)
say (Swoosh!) for (2) seconds
```

Basketball

```
when ⚑ clicked
set Player 1 Score ▼ to 0
hide
```

```
when space ▼ key pressed
set made basket ▼ to no
start sound pop ▼
go to Cat ▼
set y velocity ▼ to 24
show
repeat until  y position < -130
    change x by 8
    change y by y velocity
    change y velocity ▼ by -2
    turn ↻ 6 degrees
    if  touching Hitbox ▼ ?  and  y velocity < 0  and  made basket = no  then
        change Player 1 Score ▼ by 1
        set made basket ▼ to yes
        broadcast swoosh ▼
hide
```

CHEAT MODE: FREEZE THE HOOP

The moving basketball hoop is a hard target to hit. Let's add a cheat that will freeze it in place for a few seconds when a player presses the 7 key.

Select the Hoop sprite and click the **Make a Variable** button to create a new **For this sprite only** variable named freeze. Then modify the code for the Hoop sprite to match the following code:

```
when [flag] clicked
set freeze ▼ to no
forever
    if < freeze = no > then
        glide 1 secs to x: pick random -240 to 240 y: pick random -50 to 180
```

Remember to add the new **pick random** block for **y**!

```
when I receive swoosh ▼
start sound Cheer ▼
say Swoosh! for 2 seconds
```

```
when 7 ▼ key pressed
set freeze ▼ to yes
wait 6 seconds
set freeze ▼ to no
```

Click the green flag to test the code so far. Press the 7 key and make sure the hoop stops gliding to new positions for six seconds. Then click the red stop sign and save your program.

SUMMARY

In this chapter, you built a game that

- ▶ Implements gravity and realistic falling
- ▶ Has a side-view perspective instead of a top-down view
- ▶ Uses variables to keep track of scores, falling speeds, and the first time a basket is made
- ▶ Has a hitbox to detect when a basket is made

The use of gravity in this program was pretty simple. By the time you reach Chapter 7, you'll be able to make an advanced platformer game with more complex jumping and falling. But there are plenty of Scratch programming techniques to practice first. In Chapter 5, you'll make a side-view game that uses cloning to duplicate a sprite dozens of times.

REVIEW QUESTIONS

Try to answer the following practice questions to test what you've learned. You probably won't know all the answers off the top of your head, but you can explore the Scratch editor to figure out the answers. (The answers are also online at *http://www.nostarch.com/scratch3playground/*.)

1. How is a side-view game (like the *Basketball* game) different from a top-down game (like the *Maze Runner* game)?

2. What can a variable store?

3. What is the difference between **For this sprite only** and **For all sprites**?

4. How can you make a sprite jump?

5. When the cat jumps in the *Basketball* game, what keeps it from just going up forever?

6. What is the difference between the **glide** and **go to x y** blocks?

7. How do you make the code inside an **if then** block run only if *two* conditions are true?

5

A POLISHED BRICK BREAKER GAME

ave you ever seen a brick breaker game? The player controls a paddle at the bottom of the screen to bounce a ball that breaks blocks at the top of the screen. The player loses when the ball gets past the paddle. This game is simple enough to program, but it can look a bit boring. In this chapter, you'll learn a few tricks to make the game more colorful and interesting by adding animations and effects.

You'll use an iterative process: first, you'll make the basic game; then, you'll make small improvements to it. The result will be a more professional-looking game that other Scratchers on the Scratch website will think looks awesome.

The following figure shows what the *Brick Breaker* game looks like before and after you've polished it:

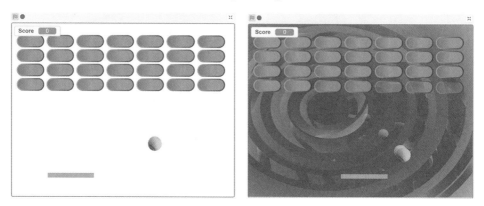

Before you start coding, take a look at the final game at *https://www.nostarch.com/scratch3playground/*.

SKETCH OUT THE DESIGN

Let's start by drawing what the game should look like. The sketch for *Brick Breaker* might look something like the following figure.

Score: 0

D Make clones of the brick.

E Make the ball bounce off bricks.

A Make a paddle that moves left and right.

C Make the ball bounce off the paddle.

B Make a ball that bounces off the walls.

F Make "You Win" and "Game Over" messages.

If you want to save time, you can start from the skeleton project file, named *brickbreaker-skeleton.sb3*, in the resources ZIP file. Go to *https://www.nostarch.com/scratch3playground/* and download the ZIP file to your computer by right-clicking the link and selecting **Save link as** or **Save target as**. Extract all the files from the ZIP file. The skeleton project file has all the sprites already loaded, so you'll only need to drag the code blocks into each sprite.

A MAKE A PADDLE THAT MOVES LEFT AND RIGHT

The player will control the paddle by moving the mouse. The ball bounces off the paddle toward the bricks, but the player loses if the ball gets past the paddle.

1. Create the Paddle Sprite

We don't need the orange cat in this game, so right-click or long press the Sprite1 cat in the Sprite List and select **delete** from the menu. Then click the **Choose a Sprite** button and select the green Paddle sprite.

Next, add this code to program the Paddle sprite to make it follow the mouse along the bottom of the Stage:

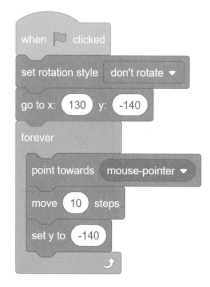

The Paddle sprite constantly moves 10 steps directly toward the mouse, but its y position stays set to -140.

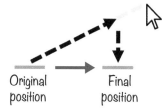

The paddle's horizontal movement is the result of two moves: moving 10 steps toward the mouse and then setting its y position to -140.

The Paddle sprite will move only left and right because its y position is always set to the bottom of the Stage (-140).

EXPLORE: ROTATION STYLES

The rotation style sets how the sprite looks when it changes direction. The three rotation styles are all around, left-right, and don't rotate. You can set this with the **set rotation style** block in the blue Motion category.

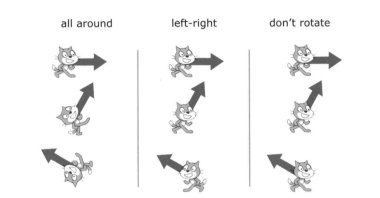

When a sprite is set to all around, it will face exactly where its direction points. But this won't work for a side-view game (like the *Basketball* game in Chapter 4), because the sprite will be upside down when its direction faces left. Instead, for these games you'll use the left-right rotation style. The sprite will face only 90 degrees (right) or −90 degrees (left), whichever is closest to the sprite's direction. If you don't want the sprite to rotate at all, even as its direction changes, set the rotation style to don't rotate.

Because we're changing the Paddle sprite's direction, we also need to set the sprite's rotation style with the **set rotation style** block. The Paddle sprite is programmed to face and move toward the mouse, but we want the sprite to always look flat and horizontal, so the rotation style is set to don't rotate.

SAVE POINT

Click the green flag to test the code so far. Move the mouse around and see if the Paddle sprite follows it. Make sure the Paddle stays at the bottom of the Stage. Then click the red stop sign and save your program.

B MAKE A BALL THAT BOUNCES OFF THE WALLS

The Scratch Sprite Library has several sprites you could use for the ball, but let's use the Tennis Ball sprite for this game.

2. Create the Tennis Ball Sprite

Click the **Choose a Sprite** button in the lower right and select the Tennis Ball sprite from the Sprite Library window. Add the following code:

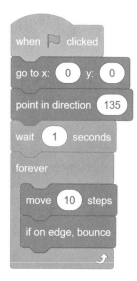

When the game starts, the Tennis Ball sprite starts at position (0, 0) in the center of the Stage; then the Tennis Ball sprite points down and to the right, toward the Paddle sprite. Next, in the **forever** loop, the Tennis Ball sprite starts to move. When the Tennis Ball sprite touches the edge of the Stage, it will bounce in a new direction.

C MAKE THE BALL BOUNCE OFF THE PADDLE

The Tennis Ball sprite now bounces off the walls but not off the Paddle sprite. Let's add that code now.

3. Add the Bounce Code to the Tennis Ball Sprite

Add the following code to the Tennis Ball sprite so it will bounce off the Paddle sprite. To do so, you'll need to create a new broadcast message, bounce.

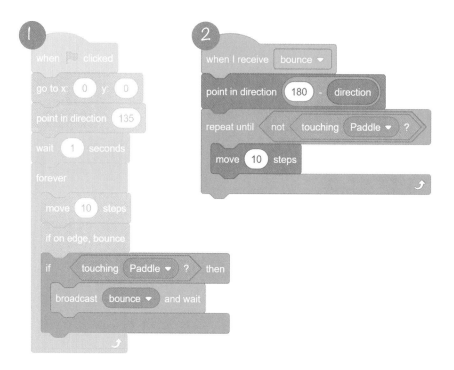

You'll use the broadcast message in script ❶ to control what happens when the ball touches the paddle in script ❷.

The **point in direction 180 - direction** code in script ❷ might seem a bit mysterious, but this equation just calculates the direction in which the ball will bounce based on the ball's current direction. If the ball is pointed up and right (45 degrees), then when it bounces off the bottom of a brick, its new direction will be down and right (135 degrees, because $180 - 45 = 135$). If the ball is pointed up and left (−45 degrees), then when it bounces off the bottom of a brick, its new direction will be down and left (225 degrees, because $180 - (−45) = 225$).

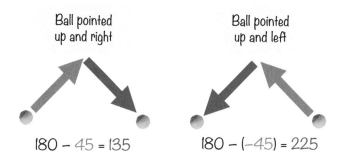

You'll use this broadcast message again later in the program when you add code to make the ball bounce off the bricks.

SAVE POINT

Click the green flag to test the code so far. Make sure the tennis ball bounces off the paddle. Then click the red stop sign and save your program.

EXPLORE: CLONING

The **create clone of myself** block makes a duplicate of the sprite, which is called a *clone*. This feature is handy whenever you want to create many copies of an object in your game, such as a lot of bad guys that look the same, a bunch of coins for the player to collect, or, in the *Brick Breaker* game, the bricks you need to hit.

Let's look at how clones work. Open Scratch in a new tab and create a new program. Add this code to the Cat sprite:

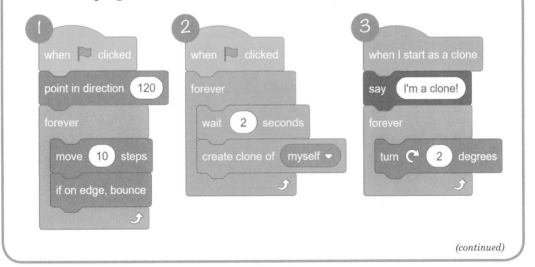

(continued)

Script ❶ makes the Cat sprite bounce around the Stage, just like the Tennis Ball sprite does in the *Brick Breaker* game. But in script ❷, we create a clone again and again, every 2 seconds. Script ❸ uses the **when I start as a clone** block to control the behavior of the cloned sprites. What do you think will happen when you run this code? Run the code now to find out if you were right.

The original Cat sprite bounces around the Stage. Every two seconds a duplicate of the sprite is created: these are the clones. Each clone will then start to turn because of script ❸.

Ⓓ MAKE CLONES OF THE BRICK

Now the game needs lots of bricks, so you'll create one Brick sprite and then clone it using Scratch's **create clone** block.

4. Add the Brick Sprite

Click the **Choose a Sprite** button in the lower right and select the Button 2 sprite from the Sprite Library window. Select Button 2 from the Sprite List and rename this sprite Brick in the Sprite Pane.

You'll have to create a new variable by selecting the orange *Variables* category and clicking the **Make a Variable** button. Name this variable Score and set it to **For all sprites**. Then add the following code to the Brick sprite:

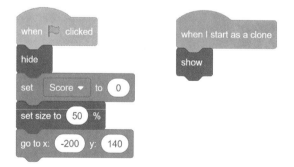

At the beginning of the game, the Score variable is set to 0 to remove any points from a previous game. The original sprite hides itself with the **hide** block, shrinks in size by 50 percent, and moves to the top-left corner of the Stage at (−200, 140). The clones, which we'll create next, show themselves with the **show** block.

5. Clone the Brick Sprite

For the *Brick Breaker* game, we want many rows of bricks. To make the rows of bricks, we'll move the original sprite across the top of the screen, creating a trail of clones. Add the following code to the Brick sprite. (Be sure not to confuse the **set x to** and **change x by** blocks!)

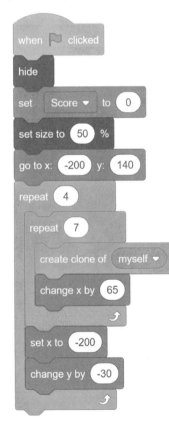

This code will create clones of the Brick sprite for all the bricks in the game, as shown here:

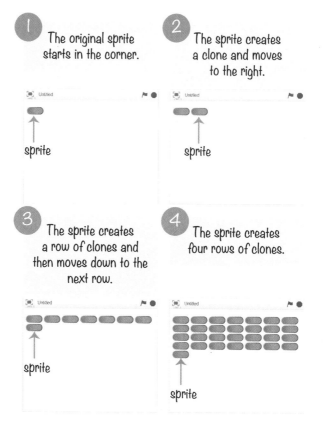

The original sprite moves to the top-left corner of the Stage at (−200, 140) ❶. Then the **repeat 7** block repeatedly moves 65 steps to the right while making clones of the sprite ❷ to create a row of seven Brick clones ❸. The **repeat 4** block repeats the row-creating code to create four rows of Brick clones ❹. Seven Brick clones multiplied by four rows results in 28 Brick clones. The 29th Brick in the previous figure is the original sprite, not a clone, and we'll hide it next.

Now all the bricks on the Stage are clones, so you don't need to duplicate the code under the **when I start as a clone** block for the original sprite.

Imagine if you duplicated the sprites instead of cloning them. Then, if you wanted to change the code, you'd have to change all 28 Brick sprites. Cloning saves you a lot of time!

E MAKE THE BALL BOUNCE OFF BRICKS

The Tennis Ball sprite bounces off the Stage edges and the Paddle sprite. Now let's make it bounce off the Brick clones.

6. Add the Bounce Code to the Brick Sprite

Update the code for the Brick sprite to match this:

```
when I start as a clone
show
forever
    if    touching   Tennis Ball ▾  ?   then
        broadcast   bounce ▾
        change   Score ▾   by   1
        delete this clone
```

When the Tennis Ball sprite hits a Brick sprite, the Brick sprite broadcasts the bounce message, which brings the Tennis Ball code into play. The ball's direction changes, just like it does when it hits the paddle. The program adds 1 to the player's Score, and then the clone deletes itself.

SAVE POINT

Click the green flag to test the code so far. Make sure the top part of the Stage fills with Brick clones and the clones disappear when the Tennis Ball sprite bounces off them. Then click the red stop sign and save your program.

Ⓕ MAKE "YOU WIN" AND "GAME OVER" MESSAGES

You need two more sprites for this game, but they won't appear until the game ends. I created mine with the Paint Editor's **Text** tool. If the player breaks all the Brick clones, the program displays the You Win sprite. If the tennis ball gets past the paddle, the program displays the Game Over sprite.

7. Modify the Tennis Ball Sprite's Code

When the Tennis Ball sprite gets past the Paddle sprite—that is, when the Tennis Ball sprite's **y position** is less than -140—the game is over. Once the game ends, the Tennis Ball sprite should broadcast a game over message. Add the following code to the Tennis Ball sprite.

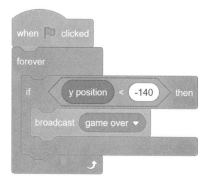

The game over broadcast will tell the Game Over sprite to appear. Let's create the sprite next.

8. Create the Game Over Sprite

Click the **Paint** button, which appears after you tap or hover over the **Choose a Sprite** button. When the Paint Editor appears, use the **Text** tool to write *GAME OVER* in red.

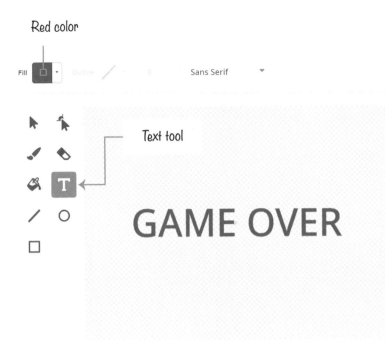

Red color

Text tool

GAME OVER

Select the sprite in the Sprite List and rename it Game Over. Then add this code to the Game Over sprite:

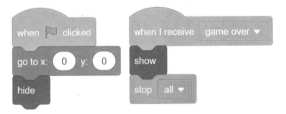

The sprite stays hidden until it receives the game over broadcast. The **stop all** block then stops all the sprites from moving as well.

9. Create the You Win Sprite

Click the **Paint** button, which appears after tapping or hovering over the **Choose a Sprite** button. In the Paint Editor, use the **Text** tool to write *You win!* in green.

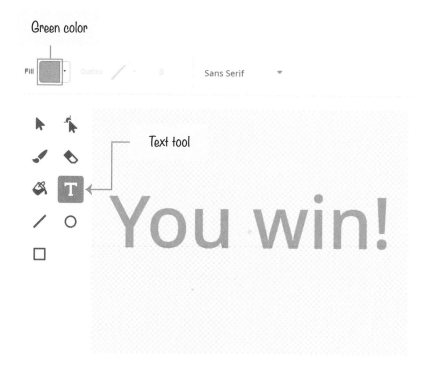

Select the sprite in the Sprite List and rename it You Win. Add this code to the You Win sprite:

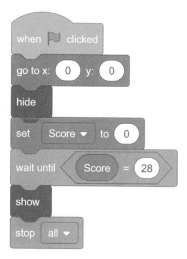

As with the Game Over sprite, the You Win sprite is hidden until a condition is met. In this game, the player needs to break all 28 bricks to win, so the condition is **Score = 28**. After the You Win sprite displays, the program stops all the other sprites from moving with **stop all**.

SAVE POINT

Click the green flag to test the code so far. Make sure the You Win sprite appears after breaking all the bricks and the program stops. To make winning the game faster, temporarily change the **wait until Score = 28** block to **wait until Score = 1**. Then you only need to break one brick to win. Save your program.

THE COMPLETE PROGRAM

The final code for the entire program is shown here. If your program isn't working right, check your code against this code.

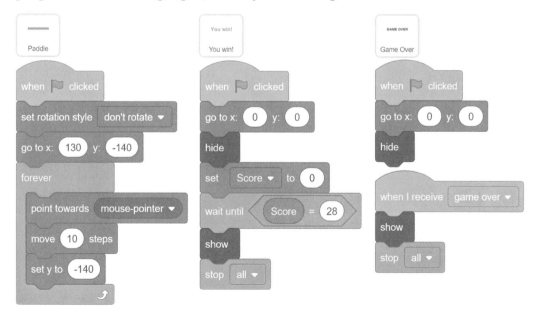

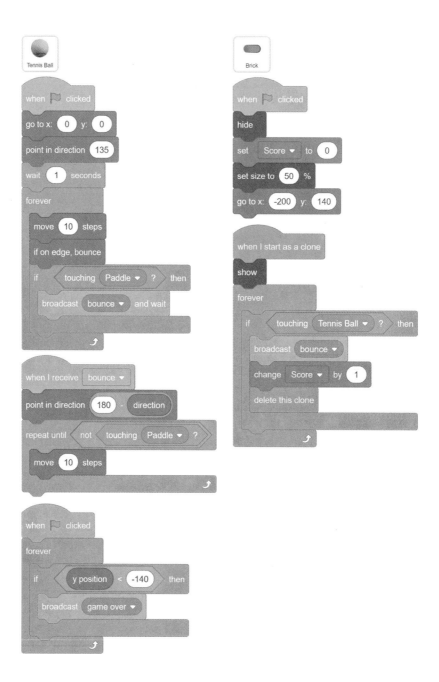

VERSION 2.0: POLISHING TIME

The game works well as it is. But now you'll add some polish. Many of the ideas for the additional features for the *Brick Breaker* game came from a Nordic Game Indie Night presentation called "Juice It or Lose It!" by Martin Jonasson and Petri Purho. The word *juice* in game design means *polish*, or modification of a game in small ways to make it feel more alive and responsive. These tricks can turn a bare-bones game into an exciting, colorful one. A juicy game looks more professional than a plain game. You can watch the presentation where Martin and Petri add juice to their *Brick Breaker* game at *https://www.nostarch.com/scratch3playground/*.

You can add many tricks to the *Brick Breaker* game to make it look polished. Even better, you can use these tricks in any of your games. Before you start coding, take a look at the complete program at *https://www.nostarch.com/scratch3playground/*.

Draw a Cool Backdrop

Adding an interesting background is a simple way to make the game look cooler. Click the **Choose a Backdrop** button and select **Neon Tunnel**. In the Paint Editor, select a color you like and use the **Fill** tool to fill in random tiles on the side of the tunnel. This can make the tunnel backdrop look more interesting.

Add Music

Sounds set the mood and make a game feel more alive. With the Stage selected in the Sprite List, click the **Sounds** tab at the top of the Block Palette. Click the **Choose a Sound** button (it looks like a speaker) in the lower left. When the Sound Library window appears, select the Dance Celebrate sound.

Then click the **Code** tab and add this code to the Stage's Code Area to give the game some background music:

Make the Paddle Flash When Hit

In this step, you'll make the paddle flash different colors when it is hit by the ball. Add the following code to the Paddle sprite:

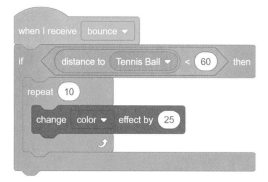

The bounce message is broadcast when the Tennis Ball sprite bounces off a Brick clone or the Paddle sprite. The **if distance to Tennis Ball < 60** block makes the color changes happen when the Tennis Ball sprite bounces off the Paddle sprite (which means the ball will be less than 60 steps away).

Click the green flag to test the code so far. Make sure the paddle flashes different colors when the ball bounces off it. Then click the red stop sign and save your program.

Add an Animated Entrance and Exit to the Bricks

The Brick clones' entrance to the game is rather boring. They just appear as soon as they're cloned. To animate their entrance, modify the Brick sprite's code to match this code.

```
when I start as a clone
change y by -10
set ghost ▼ effect to 100
show
repeat 10
    change y by 1
    change ghost ▼ effect by -10
    wait 0.01 seconds
forever
    if touching Tennis Ball ▼ ? then
        broadcast bounce ▼
        change Score ▼ by 1
        delete this clone
```

Setting **ghost effect** to 100 and then slowly decreasing it makes the Brick clones fade into view rather than instantly appear. The code also sets the Brick clones 10 steps below their final positions and slowly raises them by changing their y position in the **repeat** block. This makes the Brick clones look like they're sliding into place.

SAVE POINT

Click the green flag to test the code so far. Make sure the Brick clones fade into view rather than just instantly appearing. Then click the red stop sign and save your program.

Now let's animate the exit of the Brick clones. Modify the Brick sprite's code so the Brick clones have an animated exit instead of just instantly disappearing.

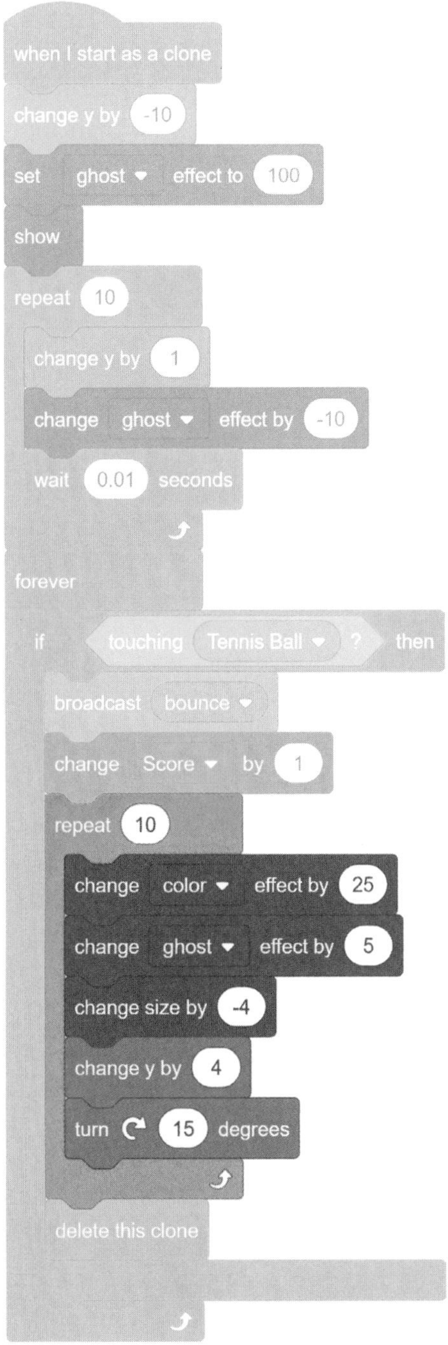

```
when I start as a clone

change y by  -10

set  ghost ▾  effect to  100

show

repeat  10
    change y by  1
    change  ghost ▾  effect by  -10
    wait  0.01  seconds

forever
    if  < touching  Tennis Ball ▾  ? >  then
        broadcast  bounce ▾
        change  Score ▾  by  1
        repeat  10
            change  color ▾  effect by  25
            change  ghost ▾  effect by  5
            change size by  -4
            change y by  4
            turn ↻  15  degrees
        delete this clone
```

Now the Brick clones will disappear in an exciting way! The **change effect by** blocks inside the **repeat** loop will make the Brick clones flash different colors and increase their ghost effect so they become more and more transparent. Meanwhile, the **change size by -4** block causes the Brick clones to shrink, the **change y by 4** block lifts them up, and the **turn clockwise 15 degrees** block rotates them. This animated exit is short but fun to watch.

SAVE POINT

Click the green flag to test the code so far. Hit the bricks with the tennis ball and make sure they spin and move up as they fade out of sight rather than just instantly disappearing. Then click the red stop sign and save your program.

Add a Sound Effect to the Brick Exit

Let's also make the Brick clones play different sound effects as they disappear. Select the Brick sprite in the Sprite List and then click the **Sounds** tab above the Block Palette. Click the **Choose a Sound** button in the lower left and select Laser1 from the Sound Library. Repeat this step to add the Laser2 sound as well.

Modify the Brick sprite's code to match this code:

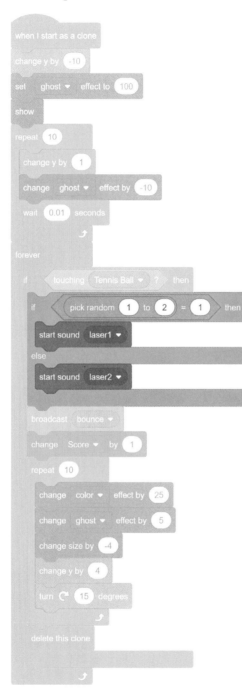

When these sounds have been loaded and the previous code has been added, the Brick clones will play a random sound effect as they disappear. The **if then else** block adds some variety to the program's sound effects by randomly selecting which sound to play. Each time the Tennis Ball sprite touches a Brick clone, the program will choose either 1 or 2 at random and then play a different sound as a result.

Add a Sound Effect to the Tennis Ball

Now you'll add a sound effect for when the Tennis Ball sprite hits the Paddle sprite. The Pop sound is already loaded for each sprite; all you have to do is update the Tennis Ball sprite's code to match this:

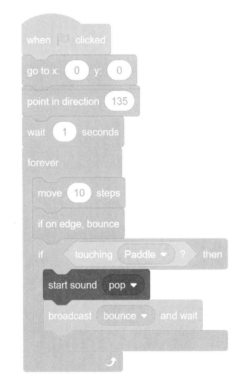

Add a Trail Behind the Tennis Ball

Adding a trail of clones behind the Tennis Ball sprite as it moves around the Stage will give it a cool comet tail. You don't want to use clones of the Tennis Ball sprite, because they would respond to the bounce broadcast whenever the original Tennis Ball sprite bounces. Instead, click the **Choose a Sprite** button in the lower right. Then select **Tennis Ball** from the Sprite Library window to create another sprite called Tennis Ball2. You'll clone this second ball to create the trail. Unlike a Tennis Ball clone, the Tennis Ball2 clones won't have a **when I receive bounce** block. Add this code to the Tennis Ball2 sprite:

```
when I start as a clone
show
go to  Tennis Ball ▼
repeat  15
    change size by  -4
    change  ghost ▼  effect by  5
delete this clone


when ⚑ clicked
hide
```

The only thing the clone is programmed to do is go to the current Tennis Ball sprite's location. The tennis ball keeps moving, but the clone stays in place, shrinking and becoming more transparent. At the end of this shrinking and fading animation, the clone is deleted.

You'll also need to update the Tennis Ball sprite with this code:

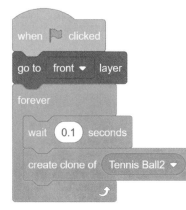

```
when ⚑ clicked
go to  front ▼  layer
forever
    wait  0.1  seconds
    create clone of  Tennis Ball2 ▼
```

This script makes a new Tennis Ball2 clone after a 0.1-second wait, creating the trail of tennis balls.

SAVE POINT

Click the green flag to test the code so far. Make sure a trail of shrinking Tennis Ball2 clones follows the original Tennis Ball sprite. Then click the red stop sign and save your program.

Add an Animated Entrance for the Game Over Sprite

When the player loses, the *GAME OVER* text simply appears. It would be more exciting if the *GAME OVER* text had an animated entrance like the Brick clones do. Modify the Game Over sprite's code to match the following code. First load the Gong sound effect by clicking the **Choose a Sound** button in the lower left after clicking the **Sounds** tab. You'll create a new broadcast message named stop game that will tell the Paddle and Tennis Ball sprites to stop moving.

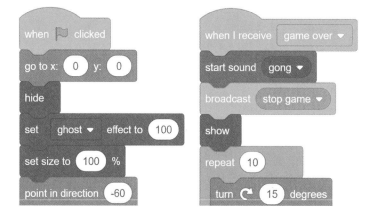

At the start of the game, the Game Over sprite hides itself and sets its ghost effect to 100. When the **show** block runs at the end of the game, the *GAME OVER* text is still completely invisible. The animation code inside the **repeat 10** block makes the *GAME OVER* text slowly fade in by changing the ghost effect by -10. The **turn clockwise 15 degrees** and **change size by 12** blocks rotate and enlarge the text. After a four-second pause, the **stop all** block ends the program.

To handle the stop game broadcast message in the Tennis Ball and Paddle sprites, add the following code to *both* of these sprites:

Add this code to the Tennis Ball and Paddle sprites.

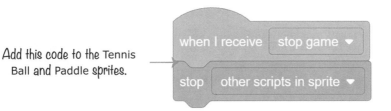

The reason you need to use the **stop other scripts in sprite** block instead of **stop all** is that the program needs to continue running while the *GAME OVER* text is animated. The **stop other scripts in sprite** block will stop the Tennis Ball and Paddle sprites from moving but let the other sprites in the program continue running. When the Game Over sprite has finished appearing on the screen, the **stop all** block will end the entire program.

SAVE POINT

Click the green flag to test the code so far. Lose the game on purpose, and make sure the *GAME OVER* text has an animated entrance instead of just instantly appearing on the screen. Then click the red stop sign and save your program.

Add an Animated Entrance for the You Win Sprite

Let's give the You Win sprite a fancy, animated entrance too. Update the code in the You Win sprite to match the following. You will have to load the Gong sound effect by clicking the **Choose a Sound** button in the lower left after clicking the **Sounds** tab.

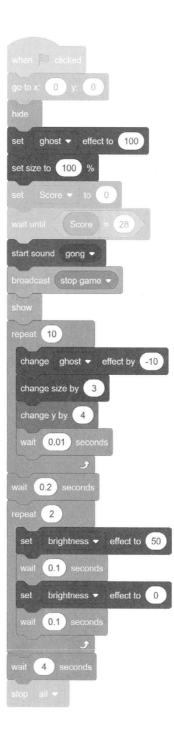

There are two sets of animations, one in the **repeat 10** block and another in the **repeat 2** block. The code in the **repeat 10** block makes the You Win sprite fade into visibility, enlarges it, and moves it upward. After this short animation plays, the **repeat 2** block's code increases the sprite's brightness to 50, waits a tenth of a second, and then resets the brightness to 0. This makes the sprite look like it's flashing. After a four-second pause, the **stop all** block ends the program.

SAVE POINT

Click the green flag to test the code so far. When you win the game, make sure the *You win!* text has an animated entrance instead of just instantly appearing on the screen. To make winning the game faster, temporarily change the **wait until Score = 28** block to **wait until Score = 1**. Then you need to break only one brick to win. Save your program.

SUMMARY

In this chapter, you built a game that

- ▶ Uses clones to quickly create many copies of the Brick sprite and a trail of Tennis Ball2 sprites
- ▶ Controls the Paddle sprite with the mouse instead of the keyboard arrow keys
- ▶ Shows the player *GAME OVER* and *You win!* messages that you created using the Paint Editor's **Text** tool
- ▶ Has several animated entrances and exits for sprites
- ▶ Uses sound effects and background music to make the game feel more alive

Making the *Brick Breaker* game provided you with several techniques that you can add to future games. You can include animated entrances, color flashes, and sound effects in many programs to make them more exciting and fun. But it's always best to make sure the plain, basic version of your game is working first and then start making it look cooler later.

This chapter also introduced cloning, which is a useful technique for creating duplicate sprites while your program runs. As you read on, the games you create will become more sophisticated, but don't worry: you just have to keep following the instructions step-by-step!

REVIEW QUESTIONS

Try to answer the following practice questions to test what you've learned. You probably won't know all the answers off the top of your head, but you can explore the Scratch editor to figure out the answers. (The answers are also online at *http://www.nostarch.com/scratch3playground/*.)

1. How does the program know when the Tennis Ball sprite has gotten past the Paddle sprite?

2. Which block creates clones of a sprite?

3. Which block has the code that clones run when they are created?

4. What are the three rotation styles?

5. Why do the You Win and Game Over sprites hide themselves after you click the green flag?

6. What does the **wait until** block do?

6
ASTEROID BREAKER . . .
IN SPACE!

steroids is a classic game developed in 1979 by Atari. Since then, many programmers have remade the game, and it's a great programming project to make in Scratch, too. The player pilots a spaceship that must destroy space asteroids while avoiding the pieces that break off . . . in space! (It's a well-known fact that adding " . . . in space!" makes everything more exciting.)

Instead of directly controlling where the spaceship moves, the player *pushes* the spaceship like a hockey puck on ice; because the player's ship has inertia, it slides around the Stage. To slow down the spaceship, players must push it in the opposite direction. It takes skill to move the spaceship without losing control, but that's half the fun of the game. The other half of the fun is blowing up asteroids.

Before you start coding, look at the final *Asteroid Breaker* program at *https://www.nostarch.com/scratch3playground/*.

SKETCH OUT THE DESIGN

Let's draw on paper what the game should look like. In our version of the game, the player controls their spaceship with the WASD keys and aims at incoming asteroids with the mouse.

Here is what my sketch looks like:

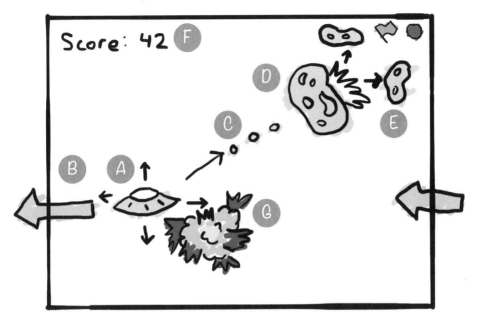

And here's what we'll be doing in each part:

A. Make a spaceship that is pushed around
B. Make the spaceship wrap around the edges
C. Aim with the mouse and fire with the spacebar
D. Make asteroids that float around
E. Make asteroids split in two when hit
F. Keep score and make a timer
G. Make the spaceship explode if it is hit

If you want to save time, you can start from the skeleton project file, named *asteroidbreaker-skeleton.sb3*, in the resources ZIP file. Go to *https://www.nostarch.com/scratch3playground/* and download the ZIP file to your computer by right-clicking the link and selecting **Save link as** or **Save target as**. Extract all the files from the ZIP file. The skeleton project file has all the sprites already loaded, so you'll only need to drag the code blocks into each sprite.

(A) MAKE A SPACESHIP THAT IS PUSHED AROUND

Before we code the exciting parts of the game, we need to set up the backdrop and sprite. We'll make this game spacey by adding the stars backdrop and the spaceship sprite. Click the **Choose a Backdrop** button in the lower right and then select **Stars**.

We won't use the Cat sprite that Scratch starts with, so right-click or long press that sprite in the Sprite Pane and select **delete** before continuing, or click the Trash Can icon next to the sprite. Start a new project in the Scratch editor and enter *Asteroid Breaker* as the project name.

1. Create the Spaceship Sprite

We'll use a flying saucer image for the spaceship, which you'll find in the resources ZIP file.

Hover over the **Choose a Sprite** button and click the **Upload Sprite** button that appears. Then select the *Spaceship.png* image file from the resources ZIP file.

In the orange *Variables* category, click **Make a Variable** and create a variable named x velocity. Make x velocity a **For this sprite only** variable. Repeat the preceding steps to create a variable named y velocity.

NOTE   *If* **For this sprite only** *does not appear, the Stage is selected instead of the* Spaceship *sprite. Close the New Variable window, select the* Spaceship *sprite, and click* **Make a Variable** *again.*

You'll also need to create two variables named Score and player is alive, but make these variables **For all sprites**.

Next, add the following code to the Spaceship sprite. The code defines the ship's starting position and the initial values of variables; it also contains the logic that defines the user's controls.

```
when [flag] clicked
go to x: (0) y: (0)
show
set [player is alive ▼] to (yes)
set [Score ▼] to (10)
set [x velocity ▼] to (0)
set [y velocity ▼] to (0)
forever
    if < key (d ▼) pressed? > then
        change [x velocity ▼] by (0.5)
    if < key (a ▼) pressed? > then
        change [x velocity ▼] by (-0.5)
    if < key (w ▼) pressed? > then
        change [y velocity ▼] by (0.5)
    if < key (s ▼) pressed? > then
        change [y velocity ▼] by (-0.5)
    change x by (x velocity)
    change y by (y velocity)
```

You might be wondering why we didn't reuse code from previous games in which we used the **change x by** or **change y by** block. In this program, holding down one of the WASD keys adds to or subtracts from the x velocity and y velocity variables. Then the code at the bottom of the script changes the x and y positions of the Spaceship sprite by using the values in these variables. Even after the player lets go of the key, the variables still reflect the updated position, so the spaceship continues to move.

Ⓑ MAKE THE SPACESHIP WRAP AROUND THE EDGES

When you tested the code, did you notice that the Spaceship sprite stops immediately when it runs into the edge of the Stage? The reason is that Scratch prevents sprites from moving off the Stage, which is helpful in most Scratch programs. But in *Asteroid Breaker*, we want sprites to go off the side of the Stage and *wrap around* to the other side.

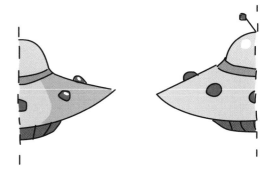

2. Add the Wrap-Around Code to the Spaceship Sprite

The following code will make the spaceship travel to the other side of the Stage whenever it reaches an edge. Add this code now.

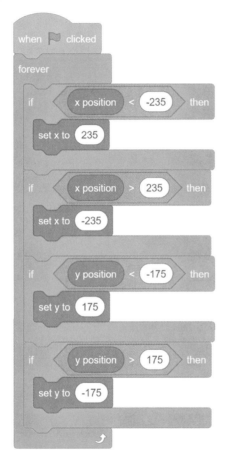

The left and right edges of the Stage are at x-coordinates −240 and 240, respectively. The bottom and top edges of the Stage are at the y-coordinates −180 and 180. We use these boundaries to write code that changes the position of the Spaceship sprite when it goes past these four coordinates. Whenever the x or y position of the Spaceship sprite gets within five steps of these edges, this new code will move the Spaceship sprite to the other side of the Stage. Because the x velocity and y velocity variables will still be moving the Spaceship sprite at the same speed and in the same direction, the Spaceship sprite will look like it's moving continuously around the Stage.

SAVE POINT

Click the green flag to test the code so far. Make sure that all four edges send the Spaceship sprite to the other side. Then click the red stop sign and save your program.

3. Add the Random-Push Code to the Spaceship Sprite

The controls for this game offer a challenge, but let's make playing the game even more difficult. We'll add random little pushes to the Spaceship sprite so that the player can't just stay in the center without moving at all.

Add the following code to the Spaceship sprite to make random pushes happen every second:

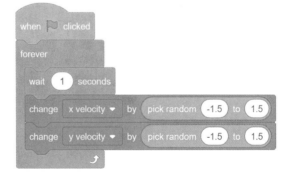

Inside the **forever** loop, the x velocity and y velocity variables are changed by a small, random amount after a one-second pause. This means that every second, the spaceship's movement receives a random push.

Click the green flag to test the code so far. Don't press any of the WASD keys. Wait to see if the spaceship starts moving a little on its own. Then click the red stop sign and save your program.

C AIM WITH THE MOUSE AND FIRE WITH THE SPACEBAR

The code that controls the Spaceship sprite is complete, so let's add energy blasts. These blasts will bust up those dangerous space asteroids! In space!

4. Create the Energy Blast Sprite

Scratch's Sprite Library has a sprite we can use for the energy blasts. Click the **Choose a Sprite** button in the lower right. Select the Ball. In the Sprite Pane, rename it Energy Blast.

We want the Energy Blast sprite to make a laser sound when the Spaceship sprite fires it. Click the **Sounds** tab above the Block Palette. Then click the **Choose a Sound** button in the lower left. Select the Laser1.

We'll make clones of the Energy Blast sprite, but the clones and original sprite will run different code. There is only one Energy Blast sprite, but the player should be able to fire many energy blasts at once. We'll create clones of the original Energy Blast sprite so that more than one energy blast can be on the Stage. The original sprite will remain hidden; all the Energy Blast sprites that appear on the Stage will be clones.

We'll use a variable named I am a clone to keep track of which is the original sprite and which are clones. Click the **Code** tab to go back to the Code Area. In the orange *Variables* category, click the **Make a Variable** button. Create a **For this sprite only** variable named I am a clone.

The original Energy Blast sprite will set this variable to no, and the clones will set it to yes. Add the following code to the Energy Blast sprite:

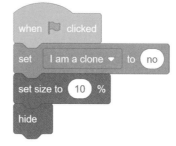

The original sprite hides itself at the start of the game and remains hidden. The clones it makes will appear on the Stage. Also, the sprite we're using is too big for the game, so set its size to 10 percent to make it smaller.

Now add the following script to the Energy Blast sprite. The player will fire an energy blast by pressing the space-bar. The original Energy Blast sprite will create clones that show themselves and move toward the mouse. Since the Energy Blast clones move toward the mouse, the player can move the mouse to aim the energy blasts.

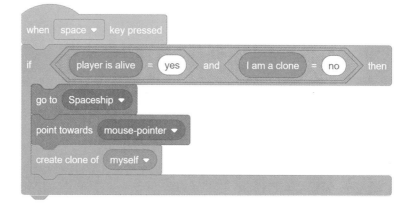

The code under the **when space key pressed** block will run for the original sprite *and* the clones if we don't give instructions that this code should only run for the original sprite. But we don't want the existing clones to create new clones. The **if then** block checks that the I am a clone variable is set to no so that only the original Energy Blast sprite runs this code and creates clones.

Obviously, the Spaceship sprite can fire Energy Blast clones only if the player is alive, so the code also checks that the player is alive variable is set to yes.

Next, add the following code to the Energy Blast sprite so that the clones move toward the mouse after they've been created:

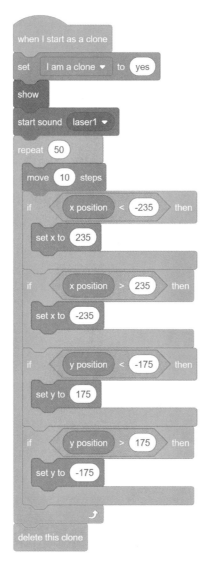

The clone starts by showing itself and moving forward. The clones should also wrap around the edges of the Stage, just like the Spaceship sprite, so we use similar code to do that here.

Notice that the clones will set their own I am a clone variable to yes. This is so that the clones do not run the code in the **when space key pressed** script. The **if then** block in that script runs the code only if **I am a clone** is set to no.

The clones move forward 10 steps at a time 50 times. That means the Energy Blast clones have a limited range and won't keep moving forever. After the loop finishes repeating 50 times, the clone deletes itself and disappears from the Stage.

SAVE POINT

Click the green flag to test the code so far. Aim with the mouse and press the spacebar to fire. Make sure the Energy Blast clones start at the Spaceship sprite and move toward the mouse. The clones should wrap around the edges of the Stage and eventually disappear. Then click the red stop sign and save your program.

D MAKE ASTEROIDS THAT FLOAT AROUND

Now we need targets for the player to hit. The asteroids in this game float around until an Energy Blast clone hits them. They'll repeatedly break apart into two smaller asteroids until they're small enough to be vaporized.

5. Create the Asteroid Sprite

Add the new asteroid sprite by hovering over the **Choose a Sprite** button, clicking the **Upload Sprite** button that appears, and selecting the *asteroid.png* file. This file is in the resources ZIP file. Click the orange *Variables* category and then click the **Make a Variable** button. Create a **For this sprite only** variable named hits. Repeat these steps to make variables named x velocity, y velocity, and rotation. All are **For this sprite only** variables. In step 6, we'll use the hits variable to keep track of how many times the asteroid has been hit and the size of the asteroid.

Let's write some code that makes the Asteroid sprite create new clones that appear on the Stage; each clone will have a random velocity and rotation. The result will be an unpredictable asteroid swarm . . . in space!

Add the following scripts to the Asteroid sprite:

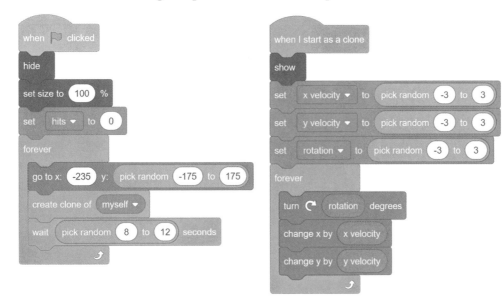

Like the Energy Blast sprite, the original Asteroid sprite will hide itself and generate clones. New clones are created every 8 to 12 seconds. When the clone is created, it shows itself, is assigned random velocities and rotations, and begins moving.

Also, like the Spaceship and Energy Blast sprites, the Asteroid sprites will wrap around the edges of the Stage. Add the following script to the Asteroid sprite:

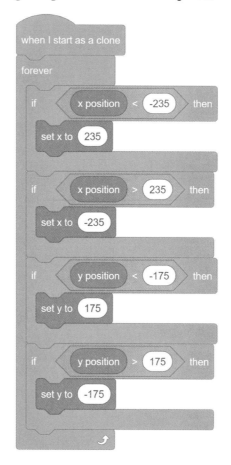

Click the green flag to test the code so far. Make sure the Asteroid sprite slowly moves and rotates and that it wraps around the edges of the Stage. Also, make sure that new clones appear every 8 to 12 seconds. Then click the red stop sign and save your program.

E MAKE ASTEROIDS SPLIT IN TWO WHEN HIT

When an Energy Blast clone hits an Asteroid, the Asteroid will create two new smaller clones of itself, making it look like the Asteroid on the Stage split in two.

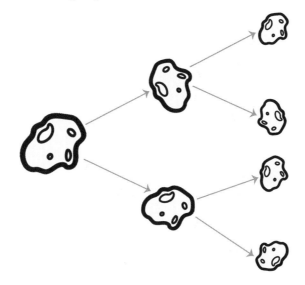

6. Add the Asteroid's Splitting Code

From the **Sounds** tab, click the **Choose a Sound** button and then select Chomp. The Chomp sound will play whenever a clone hits the asteroid.

Add the following script to the Asteroid sprite. You'll need to create a new broadcast message named asteroid blasted.

When an Energy Blast clone hits an Asteroid clone, the Asteroid clone will play the Chomp sound effect and broadcast the asteroid blasted message.

Then the Asteroid adds 2 points to the Score variable and increases the hits variable by 1. The hit Asteroid sprite will also shrink a little, changing its size by -25, so that when it (the "parent") clones itself twice, its "child" clones will be the smaller size. Finally, the Asteroid clone deletes itself.

The two smaller child clones will have hits variables that are one more than what the parent started with. The fourth time an Asteroid clone is hit, the hits variable is 4, and the code deletes the clone instead of creating two new clones. (The code after the **delete this clone** block doesn't run, because the clone no longer exists.) This prevents 1 Asteroid sprite from becoming 2, then 4, then 8, then 16, then 32, and exponentially more Asteroid sprites forever.

However, if you *do* want exponentially more Asteroid sprites, increase the number in the **if hits = 4 then** blocks.

7. Add the Asteroid Blasted Message Code to the Energy Blast Sprite

Select the Energy Blast sprite in the Sprite List and add this script to it:

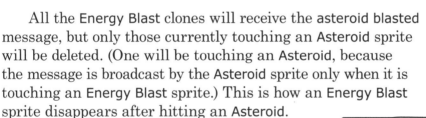

All the Energy Blast clones will receive the asteroid blasted message, but only those currently touching an Asteroid sprite will be deleted. (One will be touching an Asteroid, because the message is broadcast by the Asteroid sprite only when it is touching an Energy Blast sprite.) This is how an Energy Blast sprite disappears after hitting an Asteroid.

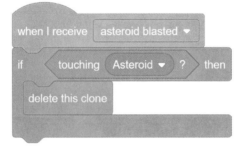

SAVE POINT

Click the green flag to test the code so far. Try blasting some of the asteroids. Make sure the energy blast disappears and the Asteroid sprite is replaced by two smaller clones. The fourth time an asteroid is hit, it should just disappear. Then click the red stop sign and save your program.

(F) KEEP SCORE AND MAKE A TIMER

The *Asteroid Breaker* game quickly becomes challenging when many tiny asteroids are flying around the Stage. A good game strategy is to be slow and careful, finishing off small Asteroid sprites before firing at larger ones. But we also want to put some pressure on the player, so let's make the Score variable start dropping by 1 point every second and have the game end when Score is 0. This way, the player will have to keep up a quicker pace when blasting Asteroid sprites.

8. Create the Out of Time Sprite

Hover over the **Choose a Sprite** button and click the **Paint** button that appears. In the Paint Editor, use the Text tool to write *OUT OF TIME* in red capital letters.

Back on the **Code** tab, go to the Sprite Pane, and rename the sprite Out of Time.

Add the following script to the Out of Time sprite:

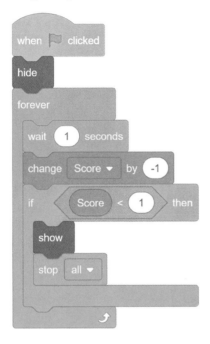

This code hides the Out of Time sprite at the start of the game and decreases the Score variable by 1 after a one-second pause. When the Score variable reaches 0, the code shows the Out of Time sprite.

SAVE POINT

Click the green flag to test the code so far. Make sure the Score variable decreases by 1 point each second. When the Score variable is 0, the Out of Time sprite should display. Then click the red stop sign and save your program.

Ⓖ MAKE THE SPACESHIP EXPLODE IF IT IS HIT

A player can lose the game if they don't blast asteroids fast enough to prevent their Score from reaching 0. But they can also lose if an asteroid hits the spaceship. Let's add code to detect when this happens and display a cool explosion animation.

9. Upload the Explosion Sprite

Eight images are available for the frames of the explosion animation. These costumes are in the *Explosion.sprite3* file in the resources ZIP file.

In the Scratch editor, hover over the **Choose a Sprite** button, click the **Upload Sprite** button that appears, and select *Explosion.sprite3*. The eight costumes for the explosion animation appear on the sprite's **Costumes** tab.

10. Add the Code for the Explosion Sprite

For the Explosion sprite, you'll create a new broadcast message named explode. When the Explosion sprite receives this message, it will appear and switch through its costumes to display the explosion animation.

The Explosion sprite will also play a sound effect when the explosion happens. Load the sound by clicking the **Sounds** tab above the Block Palette. Then click the **Choose a Sound** button in the lower left. Select the Alien Creak2 sound.

Add the following code to the Explosion sprite:

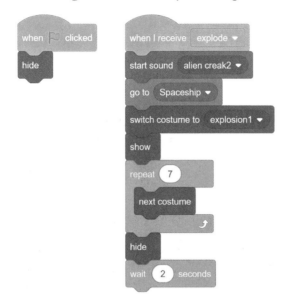

The Explosion sprite hides until it receives the explode broadcast. Then it plays a sound, goes to the position where the spaceship is, and switches its costumes seven times to create an awesome explosion animation!

11. Add the Explode Code to the Spaceship Sprite

The Spaceship sprite will broadcast the explode message when it touches one of the Asteroid clones. Add the following script to the Spaceship sprite:

The explosion animation works by briefly showing one of the costumes before moving to the next costume. This is similar to the frame-by-frame animation that cartoons and flipbooks use. Each costume is a frame, and the code quickly changes costumes to make the explosion look real.

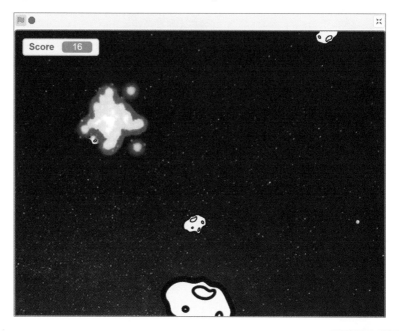

SAVE POINT

Click the green flag to test the code so far. Make sure the Explosion sprite is not visible when the game starts. Fly into an asteroid and make sure the Explosion sprite appears where the Spaceship sprite is. Then click the red stop sign and save your program.

The code for this program is too large to show here, but you can view the completed code in the resources ZIP file—the filename is *asteroidbreaker.sb3*.

VERSION 2.0: LIMITED AMMO

Once you get the hang of the game, *Asteroid Breaker* can become too easy. The one feature that makes it easy is that you can fire as fast as you can press the spacebar. This action lets the player fire indiscriminately instead of carefully aiming at the asteroids. But we can change that behavior by adding a new Energy variable. Firing an energy blast will reduce this variable by 1 each time. If the Energy variable is 0, the spaceship can't fire. The Energy variable will increase slowly over time, but it forces the player to carefully aim their shots and not waste them.

You'll need a variable to keep track of the spaceship's energy level. Select the orange *Variables* category and click the **Make a Variable** button. Make a variable named Energy and select **For all sprites**. In the Block Palette, make sure the checkbox next to Energy is checked (just like the checkbox for Score is checked) so that it will appear on the Stage.

The Energy variable will start at 10 at the beginning of the game and then decrease by 1 each time the player fires an energy blast. The player should be able to fire only if the Energy variable is greater than 0.

Modify the code in the Energy Blast sprite to match the following:

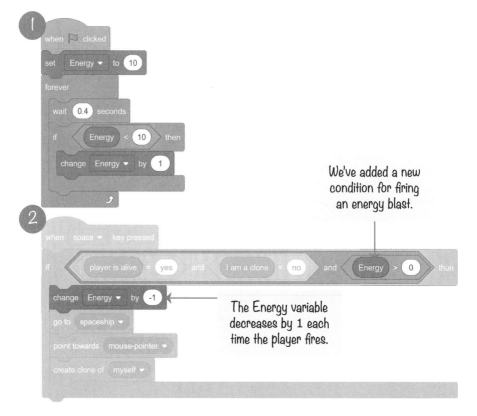

Script ❶ is new. It first sets the Energy variable to 10 before entering the **forever** loop. Inside the loop, if Energy is less than 10, the program waits for 0.4 seconds before increasing Energy by 1. This way, the Energy value never goes above 10. Script ❷ is slightly modified so that Energy must be greater than 0 for the player to fire. When an energy blast is fired, the **change Energy by -1** block decreases the Energy value.

SAVE POINT

Click the green flag to test the code so far. Make sure the Energy variable is visible on the Stage near the Score variable. The Energy variable should be 10 at the start of the game and decrease by 1 each time the player presses the spacebar. When the Energy variable is at 0, pressing the spacebar should not fire any more Energy Blast clones. Also, make sure the Energy variable recharges by 1 about every half-second. Then click the red stop sign and save your program.

CHEAT MODE: STARBURST BOMB

The limited energy in *Asteroid Breaker* version 2.0 is more challenging, but let's add a secret cheat to work around it. A cheat to have unlimited energy would be boring, so instead we'll add a special energy bomb that fires in a starburst pattern all around the spaceship.

The starburst will fire when the player presses the X key. This code is similar to the regular firing code when the player presses the spacebar.

Add the following code to the Energy Blast sprite:

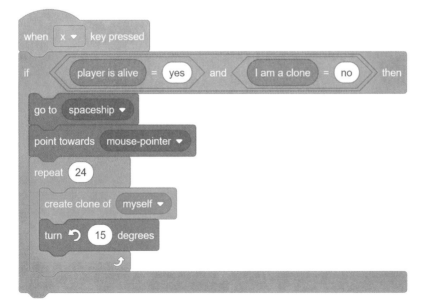

Like the **when space key pressed** script, this script checks that the player is alive and that the sprite is not a clone. Only the original sprite should run this code, not the clones.

Inside the **if then** block, the Energy Blast sprite moves to the spaceship and points in the direction of the mouse. Then, the sprite clones itself 24 times. After each clone is made, the sprite changes direction by 15 degrees counterclockwise. This results in a starburst of energy blasts in all directions.

SUMMARY

In this chapter, you built a game that does the following:

▶ Uses a hockey-puck push style to control a spaceship

▶ Has x velocity and y velocity variables to keep track of how fast the Spaceship sprite is moving

▶ Lets sprites wrap around the edges of the Stage

▶ Has Asteroid clones that can create two smaller clones of themselves

▶ Has variables Score and Energy that constantly decrease and increase, respectively, over time

▶ Has frame-by-frame animation for an explosion

This game offers players a real challenge, but as programmers, we had to add those features one by one! The player doesn't directly control the spaceship but instead pushes it. If we stopped coding the Spaceship sprite at that point, the player could just hide in a corner, safe from asteroids, so we made all the sprites wrap around the edges. Even with that addition, the player might have tried to stay still in the center of the Stage. That's when we added random small pushes to the spaceship.

It's difficult to avoid many small asteroids, so the player could have taken it slowly and carefully finished off small asteroids before targeting large ones. At that point, we made the score decrease over time to encourage the player to fire faster. The player could also keep blasting willy-nilly without aiming carefully, so we made an Energy variable to limit how fast the player could fire.

Each time you add a feature to your games, keep in mind how it affects gameplay. A game that is too hard to play is frustrating, but a game that is too easy is boring. It's all about finding a balance.

The next game is the most advanced program yet: it's a platformer game like *Super Mario Bros.* or *Super Meat Boy.* Not only will it have jumping and gravity like the *Basketball* game, but you'll be able to design custom levels for it without changing the code!

REVIEW QUESTIONS

Try to answer the following practice questions to test what you've learned. You probably won't know all the answers off the top of your head, but you can explore the Scratch editor to figure out the answers. (The answers are also online at *http://www.nostarch.com/scratch3playground/*.)

1. How does the wrap-around code work?

2. Why does the Energy Blast sprite have an I am a clone variable?

3. What prevents the Asteroid clone from breaking into exponentially more pieces forever?

4. How does the Explosion sprite's code make the spaceship look like it's exploding?

MAKING AN ADVANCED PLATFORMER

The first *Super Mario Bros.* game was introduced in 1985 and became Nintendo's greatest video game franchise and one of the most influential games of all time. Because the game involves making a character run, jump, and hop from platform to platform, this game style is called a *platformer* (or *platform* game).

In the Scratch game in this chapter, the cat will play the part of Mario or Luigi. The player can make the cat jump around a single level to collect apples while avoiding the crabs who will try to steal them. The game is timed: the player has just 45 seconds to collect as many apples as possible while trying to avoid the crabs!

Before you start coding, look at the final program at *https://www.nostarch.com/scratch3playground/*.

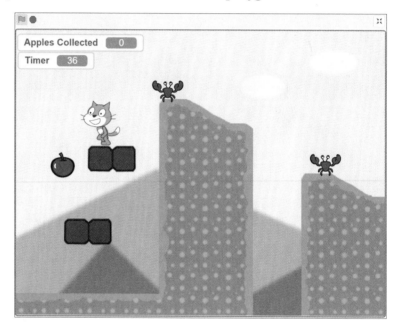

Get ready to program a more complicated game than those in previous chapters.

SKETCH OUT THE DESIGN

Let's sketch out on paper what the game should look like. The player controls a cat that jumps around while apples appear randomly. The crabs walk and jump around the platforms randomly, too.

Here's what we'll do in each part:

A. Create gravity, falling, and landing
B. Handle steep slopes and walls
C. Make the cat jump high and low
D. Add ceiling detection
E. Use a hitbox for the Cat sprite
F. Add a better walking animation
G. Create the level
H. Add crab enemies and apples

This platform game is the most ambitious one in the book, but anyone can code it if they follow the steps in this chapter. Let's code each part one step at a time.

If you want to save time, you can start from the skeleton project file, named *platformer-skeleton.sb3*, in the resources ZIP file. Go to *https://www.nostarch.com/scratch3playground/*, and download the ZIP file to your computer by right-clicking the link and selecting **Save link as** or **Save target as**. Extract all the files from the ZIP file. The skeleton project file has all the sprites already loaded, so you'll only need to drag the code blocks into each sprite.

Ⓐ CREATE GRAVITY, FALLING, AND LANDING

In the first part, we'll add gravity, falling, and landing code, similar to the *Basketball* game in Chapter 4. The important difference is that in the platform game, the cat lands when it touches a ground sprite rather than the bottom of the Stage. Coding is a bit trickier, because we want the ground to have hills and eventually platforms!

To start, click the text field at the top left of the Scratch editor and rename the project from *Untitled* to *Platformer*.

1. Create the Ground Sprite

Let's use a simple shape for the ground in the first few scripts, just to explore how the code will work.

Hover over the **Choose a Sprite** button and click the **Paint** button to create a temporary ground sprite while you learn about platforming code. In the Paint Editor, use the Brush or Line tool to draw a shape for the ground. You can make the lines thicker by increasing the number in the line width text box at the top of the Paint Editor. Be sure to draw a gentle slope on the right and a steep slope on the left.

In the Sprite Pane, rename the sprite Ground. Also, rename the Sprite1 sprite Cat.

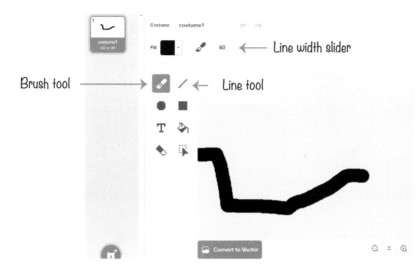

Brush tool → Line tool ←

Line width slider ←

2. Add the Gravity and Landing Code

Now that we have a sprite for the ground, we need the cat to fall and land on it.

Select the Cat sprite. In the orange *Variables* category, click the **Make a Variable** button and create a **For this sprite only** variable named y velocity. Then add the following code to the Cat sprite:

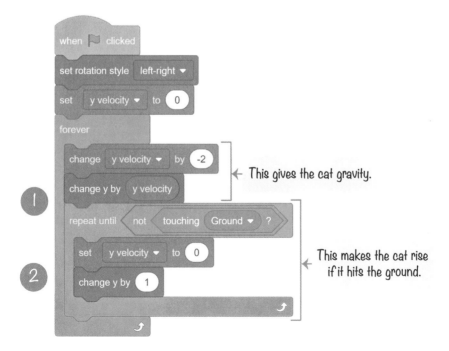

← This gives the cat gravity.

← This makes the cat rise if it hits the ground.

This code performs two actions in its **forever** loop: it makes the Cat sprite fall until it touches the Ground sprite ❶ and then lifts up the Cat sprite if it is deep in the ground ❷.

With these two code sections, the cat will fall down, hit the ground, and then rise if necessary, eventually settling on top of the Ground sprite.

In the air, falling down   Hitting the ground and rising up   Resting on top of the ground

The falling code at ❶ subtracts 2 from the y velocity variable and then moves the Cat sprite's y position by y velocity, making the cat fall faster and faster. If you programmed the *Basketball* game in Chapter 4, the falling code should be familiar.

But the **repeat until** block ❷ will loop until the Cat sprite is no longer touching the Ground sprite. (If the cat is still in the air and falling, it will not be touching the ground, so the code in the loop is skipped.) Inside this loop, the y velocity value is set to 0 so that the Cat sprite stops falling any farther. The **change y by 1** block will lift up the Cat sprite a little. The **repeat until not touching Ground** block continues lifting the sprite until it is no longer sunk into the Ground sprite. This is how the cat stays on top of the ground, no matter what the shape of the Ground sprite is.

Click the green flag to test the code so far. Drag the cat up with the mouse and let go. Make sure the cat falls and sinks into the ground a bit and then slowly lifts out of it. Then click the red stop sign and save your program.

3. Make the Cat Walk and Wrap Around the Stage

The cat also needs to walk left and right by use of the WASD keys, so add the following script to the Cat sprite:

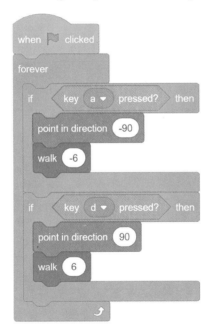

This code is straightforward: pressing A points the cat to the left (-90) and moves the x position by -6 (to the left); pressing D makes the cat point to the right and moves the x position by 6 (to the right).

Next, add the following script to make the Cat sprite wrap around to the top if it falls to the bottom of the Stage:

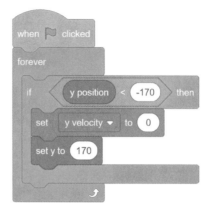

This code is similar to the wrap-around code we wrote in the *Asteroid Breaker* game in Chapter 6. We'll write wrap-around code for moving left and right later.

SAVE POINT

Click the green flag to test the code so far. Press the A and D keys to make the cat walk up and down the slopes. If the Cat sprite walks off the edge of the Ground sprite and falls to the bottom of the Stage, the Cat sprite should reappear at the top. Then click the red stop sign and save your program.

This platformer has many scripts, so you might get lost if you get confused. If your program isn't working and you can't figure out why, load the project file *platformer1.sb3* from the resources ZIP file. Click **File ▶ Load from your computer** in the Scratch editor to load the file and continue reading from this point.

4. Remove the Ground Lift Delay

The big problem with the code right now is that the Cat sprite is lifted from inside the ground to on top of it very slowly. This code needs to run so fast that the player sees the sprite only on top of the ground, not in it.

The pink custom blocks can help us do this. Go to the *My Blocks* category and click the **Make a Block** button. Name the block **handle ground** and then check the checkbox next to **Run without screen refresh**. With this box checked, Scratch will run the code in your custom pink block in Turbo Mode, even if the player didn't enable Turbo Mode by SHIFT-clicking the green flag. Otherwise, the ground-handling code would be too slow for this game.

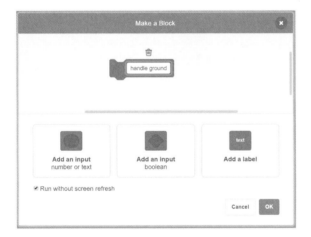

The **define handle ground** block should now appear in the Code Area. Change the code for the Cat sprite to make it use the **handle ground** block. The **handle ground** block goes where the **repeat until not touching Ground** blocks were, and that loop is moved under **define handle ground**.

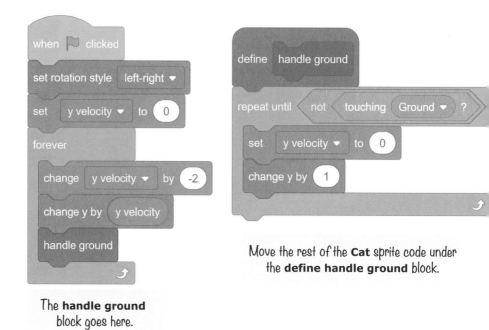

The **handle ground**
block goes here.

Move the rest of the **Cat** sprite code under
the **define handle ground** block.

This code works exactly as it did before, but now the **handle ground** block has Run without screen refresh checked, so the loop code runs in Turbo Mode. Lifting the cat now happens instantly, so it looks like the cat never sinks into the ground.

SAVE POINT

Click the green flag to test the code so far. Make the cat walk around or use the mouse to drop the cat from the top of the Stage as before. Now the Cat sprite should never sink into the ground. Then click the red stop sign and save your program.

If you're lost, open *platformer2.sb* in the resources ZIP file and continue reading from this point.

HANDLE STEEP SLOPES AND WALLS

Ⓑ

The Ground sprite has hills and slopes that the cat can walk on, and you can change the Ground sprite to pretty much any shape in the Paint Editor. This is a big improvement for the player compared to just walking on the bottom of the Stage, as in the *Basketball* game. But now the problem is that the Cat sprite can walk up the steep slope on the left as easily as it can walk up the gentle slope on the right. This isn't very realistic. We want the steep slope to block the cat. To do this, we'll make a small change to the walking code blocks.

At this point, the sprites are becoming overcrowded with lots of different scripts. So, right-click or long press the Code Area and select **Clean up Blocks** to reorganize the scripts into neat rows.

5. Add the Steep Slope Code

Now we need to edit the Cat sprite's walking code and add some new code, too. Instead of simply changing the x position by a particular value, we'll use a new custom block. Let's call it **walk** and give this new custom block an *input* called **steps**. An input is somewhat like a variable, but you can only use it in the custom block's **define** block.

Click the **Make a Block** button in the pink *My Blocks* category to make the **walk** block. Be sure to click the **Add an input number or text** button to make the **steps** input. When we want to call the new **walk** block, we'll have to define that input with a number of **steps** to take. Make sure you check the **Run without screen refresh** checkbox.

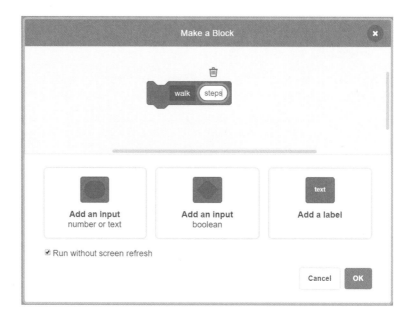

This code also requires you to make a variable for the Cat sprite named ground lift (which should be **For this sprite only**). We'll use this new variable to determine whether a slope is too steep for the cat to walk up. This code is a little complicated, but we'll walk through it step-by-step. For now, make the Cat sprite's code look like the following:

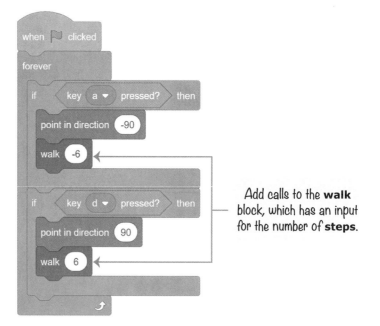

Add calls to the **walk** block, which has an input for the number of **steps**.

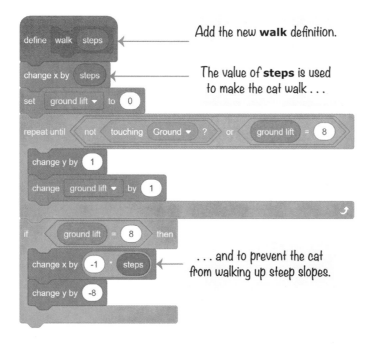

Add the new **walk** definition.

The value of **steps** is used to make the cat walk . . .

. . . and to prevent the cat from walking up steep slopes.

We want the cat to walk six units, as we did earlier, so we use -6 and 6 in the walking script when we call **walk**. In the **define walk** block, the **steps** input block is used in the **change x by** blocks. This makes the code more compact, because we can use the same script for moving the cat to the left (with the **walk -6** block) or to the right (with the **walk 6** block).

The code in the **repeat until** loop uses the ground lift variable to determine whether the slope is a walkable slope or a wall that should block the Cat sprite's progress. The ground lift variable starts at 0 and changes by 1 each time the **repeat until** loop lifts the Cat sprite's y position by 1. This loop continues looping until the sprite is no longer touching the ground or ground lift is equal to 8.

If ground lift is less than 8, then the slope isn't that steep. The sprite can walk up the slope, so the **define walk** script doesn't have to do anything else.

But if **ground lift = 8**, the **repeat until** loop stops looping. This code means "the sprite has been lifted up by 8, but it's still touching the Ground sprite, so this must be a steep slope." In that case, we need to undo the lift *and* the walking movement. The **change y by -8** and **change x by -1 * steps** blocks undo the Cat sprite's movement. Multiplying the **walk** input by -1 gives the opposite number of the input and variable.

Gentle slope: The Cat sprite is lifted eight steps and is no longer touching the ground. The Cat sprite can walk up this slope.

Steep slope: The Cat sprite is lifted eight steps but is still touching the ground. The Cat sprite cannot walk up this slope.

This code is exactly like the code in the *Maze Runner* game in Chapter 3 that blocks the player from walking through walls.

SAVE POINT

Click the green flag to test the code so far. Use the A and D keys to make the cat walk around. The cat should be able to walk up the gentle slope on the right, but the steep slope on the left should stop the cat. Click the red stop sign and save your program.

If you're lost, open *platformer3.sb* in the resources ZIP file and continue reading from this point.

ⓒ MAKE THE CAT JUMP HIGH AND LOW

With the walking code done, let's add jumping. In the *Basketball* game, we changed the falling variable to a positive number. This meant that the player jumped the height of that variable's value each time. But in many platform games, the player can do a short jump by pressing the jump button quickly or can jump higher by holding down the jump button. We want to use high and low jumping in this platform game, so we'll have to come up with something a little more advanced than the *Basketball* game's jumping code.

6. Add the Jumping Code

Let's first create a **For this sprite only** variable named in air. This variable will be set to 0 whenever the Cat sprite is on the ground. But in air will start increasing when the Cat sprite is jumping or falling. The larger the in air value is, the longer the cat will have been off the ground and in the air.

   Add this script to the Cat sprite:

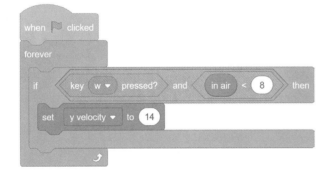

   The **forever** loop keeps checking whether the W key is being held down. If it is, this will give the Cat sprite a velocity of 14—that is, the Cat sprite will be moving up. But note that there are two conditions for the cat to continue moving up— the player must hold down W *and* the in air variable must be less than 8.

Let's edit two of the existing Cat sprite scripts to add the in air variable that limits how high the cat can jump.

When the player first holds down the W key to make the cat jump, the y velocity variable is set to 14. This makes the code in the **forever** loop in script ❶ change the Cat sprite's y position by the positive y velocity, moving it upward. At the start of the jump, the in air variable is increasing but is still less than 8. So if the player continues to hold down the W key, y velocity keeps getting set to 14 instead of decreasing because of the **change y velocity by -2** block. This causes the jump to go upward longer than if the player had held down the W key for just one iteration through the loop. But eventually in air will become equal to or greater than 8, so it won't matter if the W key is pressed. Remember that both conditions—**key w pressed** *and* **in air < 8**—must be true for the code inside the **if then** block to run.

At this point, the y velocity variable will decrease as expected, and the Cat sprite will eventually fall. In script ❷, when the cat is on the ground, the in air variable is reset to 0.

SAVE POINT

Click the green flag to test the code so far. Press the W key to jump. Quickly pressing the key should cause a small jump. Holding down the W key should cause a higher jump. Make sure the cat can jump only while it's on the ground and can't do double jumps. Then click the red stop sign and save your program.

If you're lost, open *platformer4.sb* in the resources ZIP file and continue reading from this point.

D ADD CEILING DETECTION

The cat can walk on the ground, and now walls will stop the cat from moving through them. But if the player bumps the Cat sprite's head against a platform from below, the Cat sprite will float through it! To solve this problem, we need to make some adjustments to the lifting code to add *ceiling detection*.

7. Add a Low Platform to the Ground Sprite

Add a short, low platform to the Ground sprite's costume, as shown in the following figure. Make sure it is high enough for the cat to walk under but low enough that the cat can jump up and touch it.

This platform should be low enough that the cat could hit its head on it. If it can't, then redraw the platform a bit lower.

8. Add the Ceiling Detection Code

The problem with the code is in the custom **handle ground** block. This code always assumes the Cat sprite is dropping from above and, if the Cat sprite is touching the Ground sprite, it should be lifted above it. The Ground sprite represents any solid part that the cat can't move through, including ceilings.

We need to change the code so that if the Cat sprite is jumping up when it touches the Ground sprite, the cat *stops* rising because it bumps its head. We know the Cat sprite is moving upward when its y velocity is greater than 0. So, let's edit the custom **handle ground** block to add a new Boolean input named **moving up**.

A *Boolean* is a true or false value. We use a Boolean input because we need to know whether y velocity is greater than 0 when the **handle ground** block is first called. This true or false value is stored in the **moving up** input just like a variable would store it. If we put **y velocity > 0** in the **if then** block instead of **moving up**, then the cat would end up rising above the ceiling instead of bumping against it.

Right-click or long press the **define handle ground** block and select **Edit** from the menu.

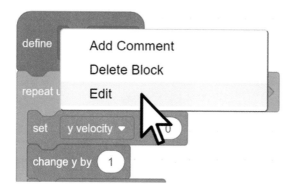

Click the **Add an input boolean** button. Name this new input field **moving up** and then click **OK**.

This will add a new **moving up** block that you can drag off the **define handle ground** block just like you do with blocks from the Block Palette. This **moving up** block will be used in a new **if then else** block. Modify the **define handle ground** block's code to match this.

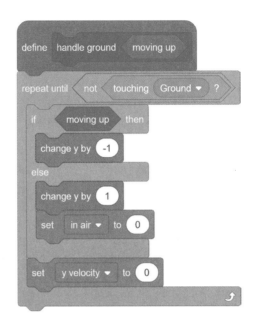

If the cat is moving up, the **change y by -1** block makes the cat look like it's bumping its head. Otherwise, the script behaves as it did previously, raising the cat so it sits above the ground.

Next, we have to edit the **handle ground** call in this script. We'll add a Boolean condition to determine whether the sprite is moving up, which is **y velocity > 0**.

The **handle ground y velocity > 0** block sets the **moving up** input to true if y velocity is greater than 0 (that is, if the sprite is jumping and moving up). If y velocity is not greater than 0, then the sprite either is falling down or is still, causing the **moving up** input to be set to false.

This is how the **define handle ground** block decides whether it should run **change y by -1** (so the Cat sprite can't go up through the ceiling) or run **change y by 1** (so the Cat sprite is lifted up out of the ground). Either way, if the Cat sprite is touching the Ground sprite (which it is if the code inside the **repeat until not touching Ground** block is running), the y velocity variable should be set to 0 so that the Cat sprite stops falling or jumping.

SAVE POINT

Click the green flag to test the code so far. Make the cat walk under the low platform and jump. Make sure the cat bumps into the platform but does not go above it. Then click the red stop sign and save your program.

If you're lost, open *platformer5.sb* in the resources ZIP file and continue reading from this point.

Ⓔ USE A HITBOX FOR THE CAT SPRITE

There's another problem with the game. Because the code relies on the Cat sprite's touching the Ground sprite, any part of the Cat sprite can be "standing" on the ground, even its whiskers or cheek! In this figure, the cat isn't falling because its cheek has "landed" on the platform, which isn't very realistic.

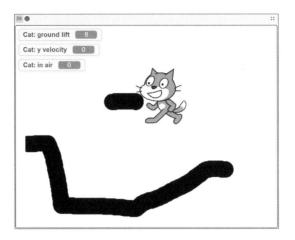

Fortunately, we can fix this problem by using the hitbox concept we used in the *Basketball* game.

9. Add a Hitbox Costume to the Cat Sprite

Click the Cat sprite's **Costumes** tab. Then hover over the **Choose a Costume** button and click the **Paint** button that appears. Draw a black rectangle that covers most (but not all) of the area of the other two costumes. The following figure shows a slightly transparent first costume in the same image so you can see how much area the black rectangle covers:

Name this costume hitbox. Whenever the *Platformer* program's code checks whether the Cat sprite is touching the Ground sprite, we'll switch the costume to the black rectangle hitbox costume before the check and then back to the regular costume after the check. By doing so, the hitbox will determine whether the cat is touching the ground.

These costume switches will be handled with the pink custom blocks that have the option Run without screen refresh checked, so the hitbox costume will never be drawn on the screen.

10. Add the Hitbox Code

We'll add **switch costume to** blocks to the start and end of both pink custom blocks. Modify the Cat sprite's code to look like this:

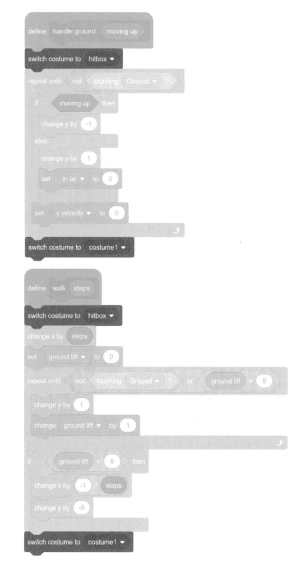

These blocks from the purple *Looks* category will change the costume to the hitbox. Because the hitbox is a simple rectangle that doesn't have protruding parts that could "catch" on platforms, like the cat's head and whiskers could, the game will behave in a more natural way.

SAVE POINT

Click the green flag to test the code so far. Make the cat jump around and make sure the cat can't hang off the platform by its cheek or tail. Then click the red stop sign and save your program.

If you're lost, open *platformer6.sb* in the resources ZIP file and continue reading from this point.

F ADD A BETTER WALKING ANIMATION

The Cat sprite that a Scratch project starts with has two costumes named costume1 and costume2.

costume1   **costume2**

You can make a simple walking animation by switching back and forth between these two costumes. However, a Scratcher named griffpatch (whose profile is at *https://scratch.mit.edu/users/griffpatch/*) has created a series of walking costumes for the Scratch cat.

In addition, griffpatch has made costumes for standing, jumping, and falling.

Using these costumes will make the *Platformer* game look more polished than using the two simple costumes that the Cat sprite comes with. We just need to add some animation code that switches between these costumes at the right time. Fortunately, griffpatch has created several cool Scratch programs using the costumes in the book resources download at *https://nostarch.com/scratch3playground/*.

11. Add the New Costumes to the Cat Sprite

To add the new costumes, you must upload the costume files into your Scratch project. You'll find the eight walking images and the standing, jumping, and falling images in the resources ZIP file. The filenames for these images are *Walk1.svg*, *Walk2.svg*, and so on, up to *Walk8.svg*, as well as *Stand.svg*, *Jump.svg*, and *Fall.svg*.

Then, in the Scratch editor, click the Cat sprite's **Costumes** tab. Hover over the Choose a Costume button in the lower left, click the **Upload Costume** button, and select *Stand.svg* to upload the file. This creates a new sprite with *Stand.svg* as its costume.

Delete the original costume1 and costume2 costumes, but keep the hitbox costume. Put the costumes in the following order (it's important that you match this order exactly):

1. Stand
2. Jump
3. Fall
4. Walk1
5. Walk2
6. Walk3
7. Walk4
8. Walk5
9. Walk6
10. Walk7
11. Walk8
12. hitbox

Each costume has not only a name (like Walk1, Jump, or Fall) but also a number. The costume number is based on the costume's order in the **Costumes** tab. For example, the costume at the top is named Stand, but it is also known as costume 1. The costume beneath it is named Jump, but it is also known as costume 2. The code we add in the next step will refer to costumes by their names and numbers.

12. Create the Set Correct Costume Block

With all these different costumes, it will be a bit tricky to determine which frame we need to show and when. We'll use the idea of animation frames: several frames shown together quickly make a moving image, just like a flip book.

To keep track of the frames, create two **For this sprite only** variables named frame and frames per costume. Then add two **set** blocks for these initial variables to the Cat sprite's **when green flag clicked** script.

```
when  clicked
set size to 30 %
set rotation style  left-right ▼
set  y velocity ▼  to 0
set  frame ▼  to 0
set  frames per costume ▼  to 4
forever
    change  y velocity ▼  by -2
    change y by  y velocity
    change  in air ▼  by 1
    handle ground  y velocity > 0
```

Now the setup is done.

As the player moves the Cat sprite left or right, we want the frame variable to increase. The frames per costume variable keeps track of how fast or slow the animation runs.

Let's edit the code in the **define walk** custom block to increase the frame variable by an amount calculated from frames per costume.

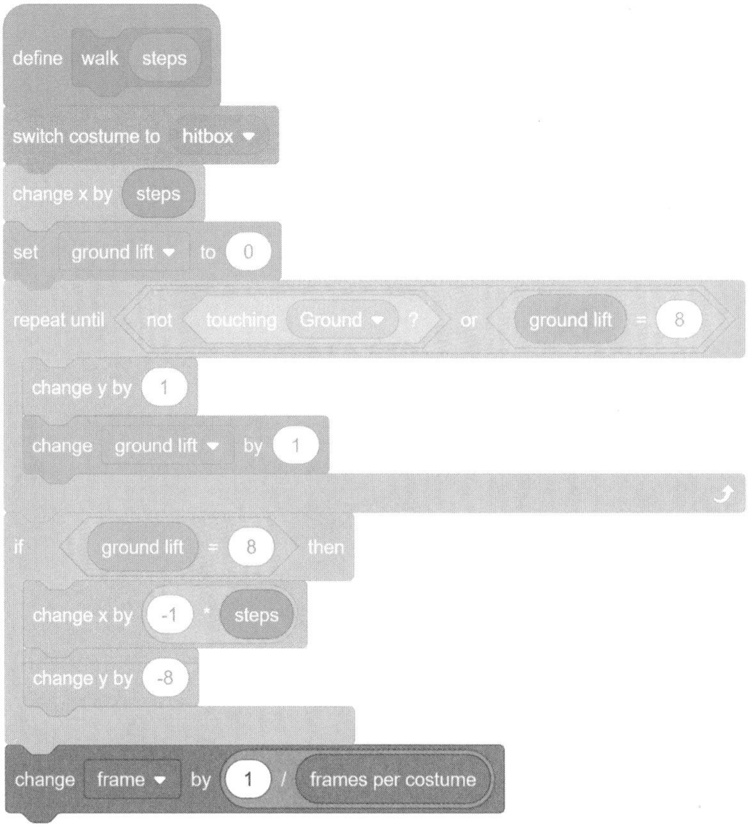

```
define walk steps

switch costume to hitbox ▼

change x by steps

set ground lift ▼ to 0

repeat until  not  touching Ground ▼ ?  or  ground lift = 8
    change y by 1
    change ground lift ▼ by 1
↻

if  ground lift = 8  then
    change x by -1 * steps
    change y by -8

change frame ▼ by 1 / frames per costume
```

When the cat is standing still (that is, not moving left or right), the frame variable should be reset to 0. Modify the Cat sprite's existing **when green flag clicked** script to add a third **if then** block that resets the frame variable.

```
when clicked

forever
    if   key  a ▾  pressed?   then
        point in direction  -90
        walk  -6

    if   key  d ▾  pressed?   then
        point in direction  90
        walk  6

    if   not  key  a ▾  pressed?   and  not  key  d ▾  pressed?   then
        set  frame ▾  to  0
```

Be sure to put the green **and** block in the **if then** block first. You want the **not** blocks to go inside the **and** block.

Now let's write some code to determine which costume to show. We'll use this code in a few places in the scripts we've written, so let's make a custom block.

In the pink *My Blocks* category, click the **Make a Block** button and name this block **set correct costume**. Check the option **Run without screen refresh** and then click **OK**. Add the following blocks to the Cat sprite, starting with the new **define set correct costume** block.

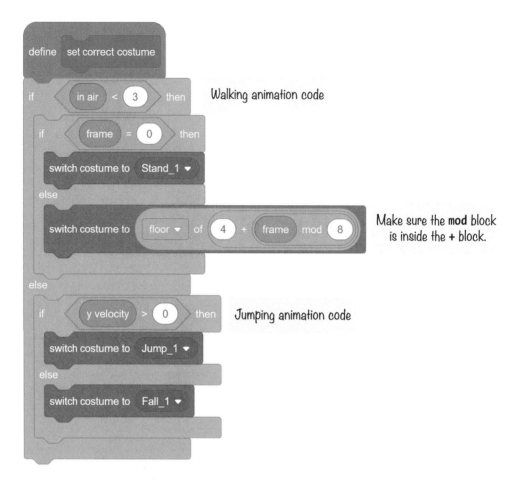

Walking animation code

Make sure the mod block is inside the + block.

Jumping animation code

If the Cat sprite is on the ground (or has just started jumping or falling so that in air is less than 3), then we want to display either the standing costume or one of the walking costumes. Remember that the **when green flag clicked** script keeps setting frame to 0 if the player isn't pressing the A key or D key. So when frame is 0, the **switch costume to Stand** block displays the Stand costume. Otherwise, we must calculate which of the eight walking costumes to show. This calculation refers to costumes by their numbers, which are based on their order in the **Costumes** tab.

Which walking costume is shown is decided by the **switch costume to floor of 4 + frame mod 8** blocks. Wow, that looks complicated! Let's break it down to better understand each part.

The **mod** block does a *modulo* mathematical operation, which is the remainder part of division. For example, 7 / 3 = 2 remainder 1, so 7 mod 3 = 1 (the remainder part). We'll use **mod** to calculate which costume number to display.

The frame variable keeps increasing, even though we have only eight walking costumes. When frame is set to a number from 0 to 7, we want it to display costumes 4 to 11. This is why our code has the **4 + frame** blocks. But when frame increases to 8, we want to go back to costume 4, not costume 12.

The **mod** block helps us do this wrap-around with the costume numbers. We can control the costume that is displayed by using a math trick: because 8 mod 8 = 0, a frame value of 8 will show the first walking costume! We have to add 4 to this number, because the first walking costume is actually costume 4. (Remember, costume 1, costume 2, and costume 3 are the standing, jumping, and falling costumes, respectively.) This sum is then used with the **floor** block. *Floor* is a programming term that means "round down." Sometimes **frame** will be set to a number like 4.25 or 4.5, so **4 + frame** would be 8.25 or 8.5, but we just want to round down to show costume 8.

Whew! That's the most math you've seen in this book so far, but when it's broken down, it becomes easier to understand.

The code in the **else** part of the **if then else** block handles what happens if in air is greater than or equal to 3. We check y velocity to see whether the cat is falling (that is, if y velocity is less than or equal to 0) or jumping (that is, if y velocity is greater than 0) and switch to the correct costume. Finally, the **define set correct costume** code finishes.

Replace the **switch costume to costume1** blocks in the **define handle ground** and **define walk** blocks with the new

set correct costume blocks. Also, add the **change frame by 1 / frames per costume** blocks so that the frame variable increases over time, as shown here:

```
define handle ground (moving up)

switch costume to hitbox ▼

repeat until < not < touching Ground ▼ ? > >
    if < moving up > then
        change y by (-1)
    else
        change y by (1)
        set in air ▼ to (0)

    set y velocity ▼ to (0)

set correct costume
```

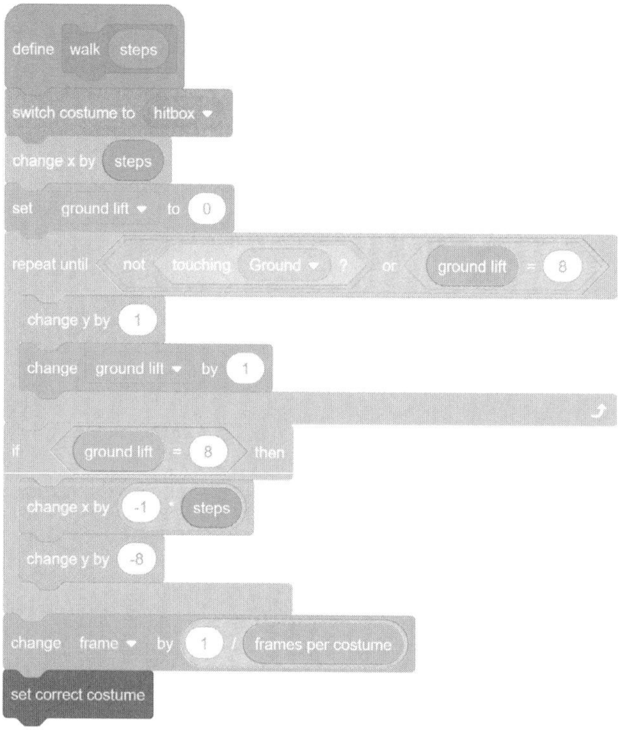

```
define walk (steps)

switch costume to hitbox ▼

change x by (steps)

set ground lift ▼ to (0)

repeat until < not < touching Ground ▼ ? > or < ground lift = 8 > >
    change y by (1)
    change ground lift ▼ by (1)

if < ground lift = 8 > then
    change x by (-1) · (steps)
    change y by (-8)

change frame ▼ by (1) / (frames per costume)

set correct costume
```

G CREATE THE LEVEL

The new walking animation makes the *Platformer* game look more appealing. Now let's change the plain white background into a real level. What's great about the code we've written for the Cat sprite to walk, jump, and fall is that it will work with a Ground sprite of any shape or color. So if we change the Ground sprite's costume (say, for different levels), we don't have to reprogram the Cat sprite!

13. Download and Add the Stage Backdrop

Click the Ground sprite's **Costumes** tab. Hover over the **Choose a Costume** button, click the **Upload Costume** button that appears, and select *PlatformerBackdrop.png*, which is in the resources ZIP file. After this costume has uploaded, you can delete the previous costume.

It's not enough to add the file *PlatformerBackdrop.png* as a costume for the Ground sprite. You must also upload it as a Stage backdrop. Hover over the **Choose a Sprite** button, click the **Upload Backdrop** button that appears, and select *PlatformerBackdrop.png* to upload it. We need to upload the file to both places because we'll be erasing all the "background parts" from the Ground sprite in the next step. We only need the Ground sprite to mark which parts the Cat sprite can walk on. The backdrop will be the image that is displayed on the Stage.

14. Create a Hitbox Costume for the Ground Sprite

The *Platformer* game code is based on the Cat sprite's touching the Ground sprite. The Ground sprite's costume is a hitbox, so if the Ground sprite's costume is a full rectangle that takes up the entire Stage, it will be treated as though the entire Stage is solid ground. We need to remove the parts of the Ground sprite's costume that are part of the background and not platforms.

The easiest way to do this is to click the Select tool in the Paint Editor. Drag a Select rectangle over the part of the costume you want to delete. After selecting an area, press DELETE to remove that piece.

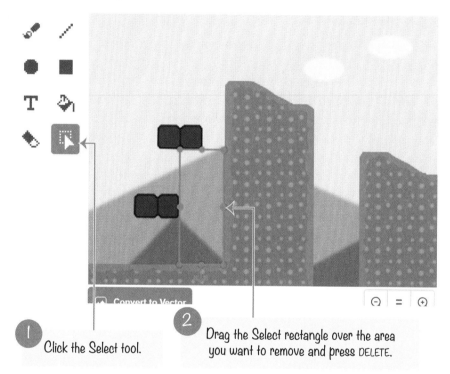

① Click the Select tool.

② Drag the Select rectangle over the area you want to remove and press DELETE.

Use the Eraser tool to erase areas that aren't rectangular. If you make a mistake, click the Undo button at the top of the Paint Editor to undo the deletion.

Keep removing the background parts of the costume until only the platform parts remain.

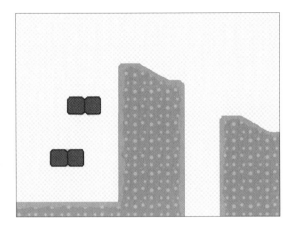

If you're having trouble creating this costume, you can use the premade *PlatformerBackdropHitbox.png* image in the resources ZIP file. The background parts of the image are already deleted, so you just need to click the **Upload Costume** button on the **Costumes** tab to add it.

15. Add the Ground Sprite's Code

The Stage backdrop is used to set the appearance of the platforms and background. The Ground sprite is used to identify which parts are solid ground that the Cat sprite can walk on. Add the following code to the Ground sprite's Code Area:

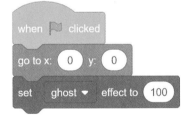

The Ground sprite's costume needs to line up perfectly over the Stage's backdrop so that you can't tell it's there. Because the Stage backdrop and the Ground sprite's costume came from the same image file, you can do this by moving the Ground sprite to the coordinates (0, 0). Otherwise, the Ground sprite's costume won't line up perfectly over the backdrop.

The *drawing* of the Ground sprite's costume doesn't matter as much as the *shape* of the costume. As long as the Ground sprite is lying perfectly on top of the backdrop, we can set the ghost effect to 100, and the Ground costume and backdrop will line up. The backdrop shows what the level looks like, while the Ground sprite acts as its hitbox.

SAVE POINT

Click the green flag to test the code so far. Make sure you can move the cat all around the Stage. Then click the red stop sign and save your program.

16. Add More Wrap-Around Code to the Cat Sprite

Notice that the level has a couple of floating platforms and a hill with a pit in the middle. When the cat falls down the pit, it wraps around the Stage and reappears at the top. Let's add wrap-around code for the left and right edges of the Stage as well. Add the following script to the Cat sprite:

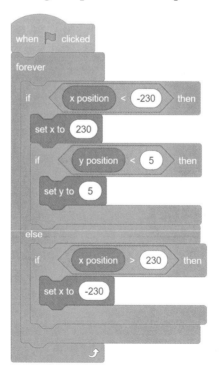

When the cat walks to the left edge of the Stage, its **x position** will be less than -230. In that case, we make it wrap around to the right edge by setting **x position** to 230.

Also, if the cat is moving from the left side of the Stage, its **y position** will be less than 5. This will put it inside the ground when moved to the right edge of the Stage, so an **if then** block checks for this condition and sets **y position** to 5.

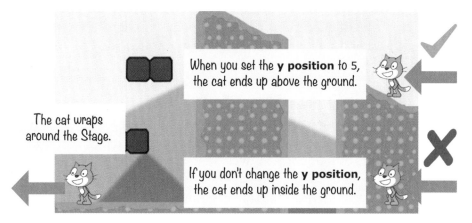

The cat wraps around the Stage.

When you set the **y position** to 5, the cat ends up above the ground.

If you don't change the **y position**, the cat ends up inside the ground.

The other **if then** block wraps the Cat to the left edge of the Stage if it's on the right edge (that is, its **x position** is greater than 230).

SAVE POINT

Click the green flag to test the code so far. Make sure the cat can walk off the left edge of the Stage and wrap around to the right, and vice versa. Then click the red stop sign and save your program.

If you're lost, open *platformer8.sb* in the resources ZIP file and continue reading from this point.

H ADD CRAB ENEMIES AND APPLES

The entire *Platformer* game setup is complete! The player can make the Cat sprite walk, jump, fall, and stand on platforms. The Cat sprite has some cool animations, and the backdrop looks like a real video game.

Now all we have to do is make a game using all the pieces we have. We'll add an apple that appears randomly around the Stage and add a few enemies that try to touch the Cat sprite and steal the apples.

17. Add the Apple Sprite and Code

Click the **Choose a Sprite** button and select the Apple sprite from the Sprite Library window that appears; then click **OK**.

As in previous games, we'll use a variable to track the game's score. Click the orange *Variables* category and then click the **Make a Variable** button to create a **For all sprites** variable named Apples Collected. This variable will keep track of the player's score.

In the Apple sprite's Code Area, add the following code:

```
when [flag] clicked
set Apples Collected ▼ to 0
set size to 50 %
forever
    set ghost ▼ effect to 100
    go to x: pick random -240 to 240 y: pick random -180 to 80
    repeat until < not < touching Ground ▼ ? >
        go to x: pick random -240 to 240 y: pick random -180 to 80
    set ghost ▼ effect to 0
    wait until < touching Cat ▼ ? >
    change Apples Collected ▼ by 1
```

At the start of the game, when the player clicks the green flag, the score in Apples Collected is set to 0. Also, because the Apple sprite is a bit too big, we set its size to 50 percent.

During the game, the Apple sprite needs to appear in random places on the Stage. We make the Apple sprite invisible using **set ghost effect to 100**. It then moves to a random place on the Stage.

But not just any random place on the Stage will do. We need to make sure the Apple sprite is not inside the Ground sprite, because it would be impossible for the player to get it. To prevent the Apple sprite from moving to somewhere inside the Ground sprite, the loop keeps trying new random places until the Apple sprite is no longer touching the Ground sprite. The movement of the apple won't be visible to the player, because the ghost effect is still set to 100. When the Apple sprite finds a place that is not touching the Ground sprite, it becomes visible again with **set ghost effect to 0**.

Then the Apple sprite waits until the Cat sprite touches it. When that happens, the score in Apples Collected increases by 1, and the Apple sprite loops to find a new random place on the Stage.

18. Create the Crab Sprite

The game wouldn't be very challenging if all the player had to do was jump around and collect apples. Let's add some enemies that the player must avoid.

Right-click or long press the Cat sprite and select **Duplicate**. The enemies will use the same code as the Cat sprite so that they can jump and fall on the platforms. (We'll remove the code that assigns the keyboard keys to control the sprite and replace it with code that moves the crabs around randomly.) Rename this duplicated sprite Crab. Then, on the **Costumes** tab, click the **Choose Costume** button and select the crab-a costume. Then open the Costume Library again and select the crab-b costume.

The Crab sprite will still have all the Cat sprite costumes (Stand, Fall, Walk1, and so on). Delete these Cat sprite costumes, including the hitbox costume. Create a new hitbox costume that is the right size for the crab. Here's what the hitbox costume should look like (the crab-a costume has been placed over it so you can see the relative size, but the costume is just the black rectangle):

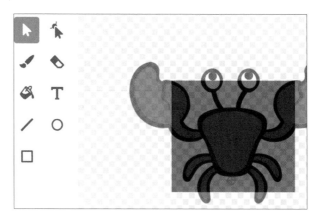

19. Create the Enemy Artificial Intelligence

In games, *artificial intelligence (AI)* refers to the code that controls the enemies' movements and how the enemies react to the player. In the platform game, the AI for the crabs is actually pretty *un*intelligent: the crabs will just move around randomly.

In the orange *Variables* category, click the **Make a Variable** button and create a **For this sprite only** variable named movement. The movement variable will store a number representing the Crab sprite's movements.

▶ Walk left
▶ Walk right
▶ Jump up straight

- ▶ Jump to the left
- ▶ Jump to the right
- ▶ Stay still

The Crab sprite's movements will be randomly decided and will frequently change. Add the following code to the Crab sprite:

```
when ⚑ clicked
forever
    set movement ▾ to (pick random (1) to (6))
    if < movement = (1) > then
        set movement ▾ to (left)
    if < movement = (2) > then
        set movement ▾ to (right)
    if < movement = (3) > then
        set movement ▾ to (jump)
    if < movement = (4) > then
        set movement ▾ to (left-jump)
    if < movement = (5) > then
        set movement ▾ to (right-jump)
    if < movement = (6) > then
        set movement ▾ to (still)
    wait (pick random (0.2) to (0.8)) seconds
```

The Crab sprite will wait a random amount of time between 0.2 and 0.8 seconds before deciding on a new random move to make.

At first, the movement variable is set to a random number between 1 and 6 that decides which movement the crab will make.

The rest of the Crab sprite's code defines these movements. Find any code from the Cat sprite code that uses the **when key pressed** blocks and replace those blocks with ones that check the movement variable. (If you ran the program right now, the keyboard keys would control the Cat *and* Crab sprites, because they have the same code!)

Modify the Crab sprite script that checks whether the player is pressing the A key or the D key to match the following code:

```
when clicked
forever
    if  movement = left  or  movement = left-jump  then
        point in direction -90
        walk -4

    if  movement = right  or  movement = right-jump  then
        point in direction 90
        walk 4

    if  movement = still  then
        set frame ▼ to 0
```

As with the Cat sprite, this code lets the Crab sprite walk left and right. Change the values in the **walk** blocks to -4 and 4 to make the crab move slower than the player.

Then change the script that handles the player pressing the W key to jump to match the following code:

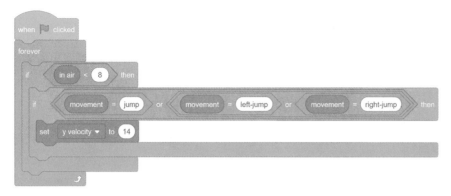

This code lets the Crab sprite jump up, left, and right.

Now let's animate the crab's movements. The Crab sprite has only two costumes, crab-a and crab-b. We'll switch between these two costumes to make it look like the crab is walking. We can simplify the **set correct costume** block for the Crab sprite a bit.

Modify the **set correct costume** code to look like the following:

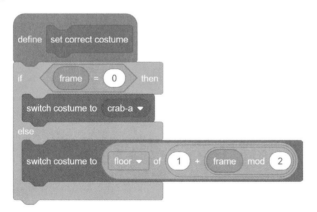

Notice that the numbers in the **floor of 1 + frame mod 2** blocks have also changed. The first costume is costume 1, and the Crab has only two costumes, so the numbers in these blocks have been changed to 1 and 2.

Finally, we need to create a new script so that the crabs can steal apples from the player. Add this code to the Crab sprite:

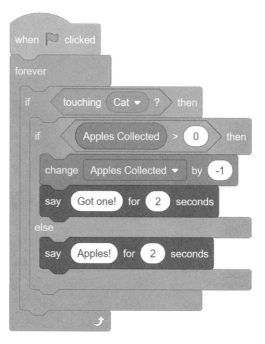

The Crab sprites subtract 1 from Apples Collected and say "Got one!" when they touch the player. If the player has 0 apples, the Crab sprites will say "Apples!" but will not subtract 1 from Apples Collected.

The game will be a bit more exciting with two crabs, so right-click or long press the Crab sprite in the Sprite List and select **Duplicate** from the menu.

SAVE POINT

Click the green flag to test the code so far. Make sure the two crabs are jumping around. When they touch the cat, they should steal an apple and say "Got one!" If the cat doesn't have any apples, the crabs should just say "Apples!" Then click the red stop sign and save your program.

20. Add the Time's Up Sprite

We're almost done! The last thing we need to add to the game is a timer. Now the player will be under pressure to grab apples as quickly as possible instead of playing it safe. Hover over the **Choose a Sprite** button, click the **Paint** button, and draw the text *Time's Up* in the Paint Editor. Mine looks like this:

Rename the sprite Time's Up. Then create a **For all sprites** variable named Timer and add the following code:

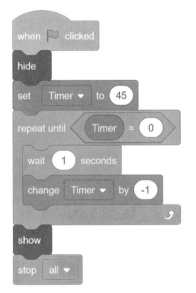

This code gives the player 45 seconds to collect as many apples as possible while trying to avoid the crabs who will steal one. When the Timer variable reaches 0, the Time's Up sprite will appear, and the game will end.

Now the *Platformer* game is ready for final testing!

SAVE POINT

Click the green flag to test the code so far. Walk and jump around, collecting apples while trying to avoid the crabs. Make sure that when the timer reaches 0, the game ends. Then click the red stop sign and save your program.

The complete code for this program is too large to list in this book. However, you can view the completed code in the resources ZIP file—the filename is *platformer.sb3*.

SUMMARY

You did it! You're a Scratch master! The advanced *Platformer* game is the most elaborate and complex project in this book. You combined and used lots of different concepts to make this game, so it might help to read through this chapter a few more times.

In this chapter, you built a game that

▶ Uses a ground sprite that the player stands on

▶ Uses pink custom blocks with the Run without screen refresh option enabled

▶ Lets the player walk up and down slopes

▶ Has ceiling detection so the player bumps their head on low platforms

▶ Has detailed animations for walking, jumping, and falling

▶ Implements AI for enemies so they move around on their own

That wraps it up for this book, but don't let that stop you from continuing your programming adventure. You can always look through other Scratchers' programs to get more ideas. Find a game you like and try to create it from *scratch*. (All my puns are intended.)

The great thing about Scratch is that it provides you with unlimited possibilities for the types of games you can make. You can create clones of popular classic games like *Pac-Man* or *Flappy Bird*. Or you can make unique games using your own designs. Good luck!

REVIEW QUESTIONS

Try to answer the following practice questions to test what you've learned. You probably won't know all the answers off the top of your head, but you can explore the Scratch editor to figure out the answers. (The answers are also online at *http://www.nostarch.com/scratch3playground/*.)

1. The pink custom blocks help you avoid duplicating code. Why is this a good thing?

2. How is a pink custom block's input like a variable?

3. Where can you use a pink custom block's input?

4. What does *modulo* mean in mathematics?

5. What does *floor* mean in programming?

WHERE TO GO FROM HERE

Ready for more? There are plenty of Scratch resources to keep things exciting.

▶ **Scratch Programming Playground Studio** (*http://www.inventwithscratch.com/studio/*) is a place to share your own games based on the ones in the book! You can check out other readers' projects while you're there.

▶ **The Scratch Forums** (*http://scratch.mit.edu/discuss/*) is a place for Scratchers to share ideas and ask and answer questions.

▶ **25 Scratch 3 Games for Kids** by Max Wainewright (No Starch Press, 2019; *https://www.nostarch.com/25scratchgames/*) is a book with more projects to build in Scratch.

▶ **ScratchEd** (*http://scratched.gse.harvard.edu/*) is an online community created for teachers and other educators who use Scratch. Share your success stories, exchange Scratch resources, ask questions, and more.

There are many fun games and animations you can make with Scratch, but it does have some limitations. Your Scratch programs might not look like the "real" games you play on a PC, game console, or smartphone.

So, it's only natural that you might want to learn how to write code in a professional programming language. There are many languages to choose from, but I recommend Python or JavaScript. Python is the easiest language to learn (aside from Scratch), but it is still a language used by professional software developers. JavaScript is not quite as easy, but it's the language used for web apps that run in your browser.

If you want to learn Python, I recommend a book I wrote: *Invent Your Own Computer Games with Python*. This book is free to read online at *https://inventwithpython.com/*, or you can buy it at *https://www.nostarch.com/inventwithpython/*. If you want to learn JavaScript, I recommend Nick Morgan's *JavaScript for Kids* (No Starch Press, 2014; *https://www.nostarch.com/javascriptforkids/*). These books are great for the next step of your journey to become a master programmer.

INDEX